How to Successfully Publish a Manuscript

Islam Mohammad Shehata
Omar Viswanath
Editors

How to Successfully Publish a Manuscript

A Step-by-Step Guide

Editors
Islam Mohammad Shehata
Department of Anesthesiology,
Faculty of Medicine
Ain Shams University Cairo
Cairo, Egypt

Omar Viswanath
Department of Anesthesiology
Creighton University School of Medicine
Phoenix, AZ, USA

ISBN 978-3-031-92537-5   ISBN 978-3-031-92538-2 (eBook)
https://doi.org/10.1007/978-3-031-92538-2

© The Editor(s) (if applicable) and The Author(s), under exclusive license to Springer Nature Switzerland AG 2025

This work is subject to copyright. All rights are solely and exclusively licensed by the Publisher, whether the whole or part of the material is concerned, specifically the rights of translation, reprinting, reuse of illustrations, recitation, broadcasting, reproduction on microfilms or in any other physical way, and transmission or information storage and retrieval, electronic adaptation, computer software, or by similar or dissimilar methodology now known or hereafter developed.

The use of general descriptive names, registered names, trademarks, service marks, etc. in this publication does not imply, even in the absence of a specific statement, that such names are exempt from the relevant protective laws and regulations and therefore free for general use.

The publisher, the authors and the editors are safe to assume that the advice and information in this book are believed to be true and accurate at the date of publication. Neither the publisher nor the authors or the editors give a warranty, expressed or implied, with respect to the material contained herein or for any errors or omissions that may have been made. The publisher remains neutral with regard to jurisdictional claims in published maps and institutional affiliations.

This Springer imprint is published by the registered company Springer Nature Switzerland AG
The registered company address is: Gewerbestrasse 11, 6330 Cham, Switzerland

If disposing of this product, please recycle the paper.

To my father, who raised me to be something
To my mom, who taught me how to be a good thing
To my family, for whom I'd like to be a great thing
I especially thank Prof. Hamdy Awad who told me that I can, and Prof. Alan Kaye who has helped me to do what I can!

Islam Mohammad Shehata

16 Things this Book Will Help You Achieve

Validate your ideas
★
Create a compelling Title
★
Compose an interesting Introduction
★
Craft a competent Discussion
★
Master the Artwork
★
Simplify Referencing
★
Write a well-structured Abstract
★
Find an appropriate Journal
★
Submit like a professional author
★
Respond to reviewers comprehensively
★
Manage Funding challenges
★
Use AI tools like a pro
★
Build a great team
★
Report publishable cases
★
Conduct systematic review and meta-analysis
★
Avoid ethical violations

Introduction

To our readers,

There are very few books on the market for those looking for help in successfully writing a scientific paper. No matter what country you may come from, clinical research is becoming a necessary part of the clinical knowledge and learning that is normally associated with medical graduate studies. With these students and medical professionals in mind, we believe this new, step-by-step guide, including both print and digital content, will enable those in the medical field to achieve similar success in the clinical research world as they progress through their studies and training. This book includes many easy-to-follow tables and figures, and it is written in an easily digestible format for both novices and experienced practitioners and researchers.

The goal of *How to Successfully Publish a Manuscript: A Step-by-Step Guide* is to provide practical guidance for novice researchers to allow them to follow a systematic guide to achieving publication success from experts in the field.

For professional authors, you will find many tips and tricks to perfect your work and increase the efficiency of the whole writing and publication process

Step by step
Bit by bit
Brick by brick

The book guides you with a full chapter for every single step required to start with a research topic and end with a formal manuscript accepted for publication. We have also included basic information about AI and practical ways to use it in every chapter. We hope that would help change the way research is done by making complex tasks faster and more efficient, allowing you to focus on the core of your project.

Unlike other books, one does not have to read the whole book to start; it is intentionally designed to provide ongoing benefits as you are reading it. Therefore, we recommend that the author start writing their own manuscript while reading our chapters.

Thank you,
Islam Mohammad Shehata, MD/PhD
Omar Viswanath, MD

Contents

1. **Ethics of Research** 1
   Rodayna Wael Elsayed, Ahmed Atef Ahmed,
   Islam Mohammad Shehata, Omar Viswanath, and Randy Richardson

2. **How to Find a Meaningful Research Question** 13
   Neveen A. Kohaf, Esraa Y. Salem, Islam Mohammad Shehata,
   Omar Viswanath, and Alan David Kaye

3. **Title and Keywords** 23
   Alaa Mohamed Ibrahim, Islam Mohammad Shehata,
   Omar Viswanath, and Latha Ganti

4. **Introduction** 31
   Esraa Elbanna, Marina Ramzy, Islam Mohammad Shehata,
   Omar Viswanath, and Farnad Imani

5. **Discussion** 47
   Alaa Abdeltawab Abouammar, Rodaina Ehab Ashour,
   Islam Mohammad Shehata, Omar Viswanath, and Natalie Strand

6. **Conclusion and Abstract** 59
   Tabia Imtiyaz Khan, Islam Mohammad Shehata, Omar Viswanath,
   and Giustino Varrassi

7. **The Artwork** 75
   Ro'a Azzam, Islam Mohammad Shehata, Omar Viswanath,
   and Shaleen Vira

8. **Referencing** 103
   Dania Imtiyaz Khan, Islam Mohammad Shehata, Omar Viswanath,
   and Jamal Hasoon

9. **How to Choose the Right Journal** 121
   Eman Hamdy Oweiss, Islam Mohammad Shehata, Omar Viswanath,
   and Rory Murphy

10	**How to Properly Submit Your Paper**	137
	Abdullah Olimy, Mohamed Ahmed Ali, Islam Mohammad Shehata, Omar Viswanath, and Robert Ravinsky	
11	**How to Raise Funds for Publication Fees?**	153
	Salwa Khaled Ahmed, Eman Hamdy Oweiss, Islam Mohammad Shehata, Omar Viswanath, and Alberto Pasqualucci	
12	**Peer Review: How to Reply to Reviewers**	161
	Mennatallah Alashker, Tasnim Awad, Islam Mohammad Shehata, Omar Viswanath, and Musa Aner	
13	**How to Write A Case Report**	183
	Farah Mohamed Ismail, Ahmed Hashim, Islam Mohammad Shehata, Omar Viswanath, and Naum Shaparin	
14	**Systematic Review: A Comprehensive Guide**	191
	Manar Ahmed Kamal, Batoul Mohamed Alaswad, Islam Mohammad Shehata, Omar Viswanath, and Sarang S. Koushik	
15	**Meta-analysis Explanation and Guidance**	205
	Ahmed Saad Elsaeidy, Reem Sayad, Rahma Sameh Shaheen, Ahmed M. Kedwany, and Cyrus Yazdi	
16	**How to Recruit a Research Team?**	253
	Mariam Elgabry, Islam Mohammad Shehata, Omar Viswanath, and Ivan Urits	
17	**AI in Research**	265
	Hamada Hamdy Elbana, Moataz Maher Emara, Abdelrahman M. Saad, Islam Mohammad Shehata, Omar Viswanath, and Mohamed Rehman	

Index 277

About the Editors

Islam Mohammad Shehata, MD, PhD is a cardiac anesthesiologist and critical care clinical consultant practicing in Cairo, Egypt.

He was born and raised in Kom Hamada city, El Beheira Governorate, which is a costal governorate in Egypt. He obtained his bachelor's degree in medicine, and master's and doctoral degrees in anesthesiology from the Faculty of Medicine, Ain Shams University, Cairo, Egypt. He then completed his visiting scholarship in cardiac anesthesia at Ohio State University Wexner Medical Center, Columbus, Ohio. He also obtained the European Diploma in Anaesthesiology and Intensive Care examination.Currently, he is a lecturer of anesthesiology in anesthesia department at Ain Shams University, and a research assistant at the Faculty of Medicine, Modern University for Technology and Information, Cairo, Egypt. He currently serves as an associate editor in the anesthesia section of *Annals of Medicine* journal and a reviewer for many well-known journals. His relevant links are as follows:
http://linkedin.com/in/islam-mohammad-shehata-00463a186
https://scholar.google.com/citations?user=tIbA7HQAAAAJ&hl=en

Omar Viswanath, MD is Double Board-Certified Anesthesiologist and Interventional Pain Medicine Physician practicing in Phoenix, Arizona. He was born and raised in Naperville, Illinois, which is a suburb of Chicago. He obtained his bachelor's degree in biology from Saint Louis University in St. Louis, Missouri, and his Medical Degree from Creighton University School of Medicine in Omaha, Nebraska. He then completed his anesthesiology residency at Mt. Sinai Medical Center of Florida in Miami Beach, Florida, where he also served as chief resident, and his interventional pain medicine fellowship at Harvard-Beth Israel Deaconess Medical Center in Boston, Massachusetts.

He is a Clinical Professor of Anesthesiology at Creighton University School of Medicine, Clinical Professor of Anesthesiology at LSU Health Sciences School of Medicine, and Clinical Assistant Professor of Anesthesiology at University of Arizona College of Medicine-Phoenix. He currently has over 450 PubMed indexed research article publications, and he serves on the Editorial Board of several prestigious peer-reviewed anesthesiology and pain medicine journals.

Ethics of Research

Rodayna Wael Elsayed, Ahmed Atef Ahmed, Islam Mohammad Shehata, Omar Viswanath, and Randy Richardson

> *Ethics is knowing the difference between what you have a right to do and what is right to do.*
>
> —Potter Stewart

1.1 Definition

A set of ethical and moral principles guiding research practices [1].

1.2 Role

- It ensures that researchers act ethically with their study participants. This involves treating them with dignity and respect, which should foster trust and confidence in the integrity of the research and data it yields, which can ultimately guide, shape, and change clinical practice.

R. W. Elsayed (✉) · A. A. Ahmed
Faculty of Medicine, Modern University for Technology and Information, Cairo, Egypt
e-mail: rodaina.94618@medicine.mti.edu.eg; ahmed16008@stemegypt.edu.eg

I. M. Shehata
Department of Anesthesiology, Faculty of Medicine, Ain Shams University Cairo, Cairo, Egypt
e-mail: islam.shehata@med.asu.edu.eg

O. Viswanath
Department of Anesthesiology, Creighton University School of Medicine, Phoenix, AZ, USA

Mountain View Headache and Spine Institute, Phoenix, AZ, USA

R. Richardson
Department of Radiology, Creighton University School of Medicine, Phoenix, AZ, USA
e-mail: randyrichardson@creighton.edu

© The Author(s), under exclusive license to Springer Nature Switzerland AG 2025
I. M. Shehata, O. Viswanath (eds.), *How to Successfully Publish a Manuscript*,
https://doi.org/10.1007/978-3-031-92538-2_1

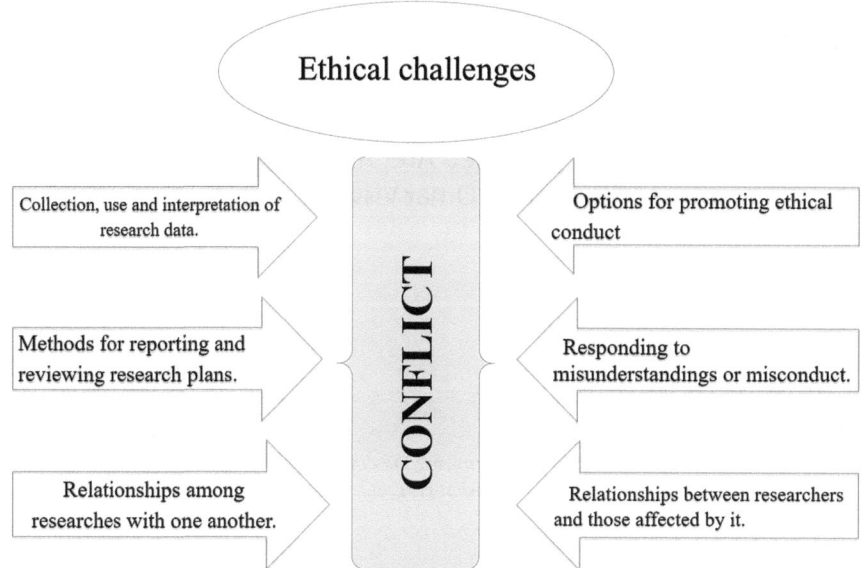

Fig. 1.1 Ethical challenges leading to conflict in research. [1]

- It regulates the relationships between the authors, peer reviewers, and journals.
- It ensures that the literature contributes positively to the community and does not violate any rights [1].

Figure 1.1 presents ethical challenges encountered in research.

1.3 History

1.3.1 Ancient Egypt and the Foundations of Medical Ethics (Oath of Imhotep)

One of the earliest known codes of medical ethics was the Oath of Imhotep, named after the renowned Egyptian physician and architect who lived around 2600 BCE. The Oath laid out the fundamental principles of medical practice, including the duty to protect the confidentiality of patients, the obligation to provide care to the best of one's abilities, and the prohibition against engaging in harmful or unethical practices.

Muslim scholars and physicians built upon the foundation laid by ancient Greek thinkers, such as Hippocrates and Galen, and further expanded the ethical principles governing the practice of medicine.

The concept of the "sanctity of life" was central to medical ethics, with physicians being required to uphold the principle of preserving life and avoiding harm.

1 Ethics of Research

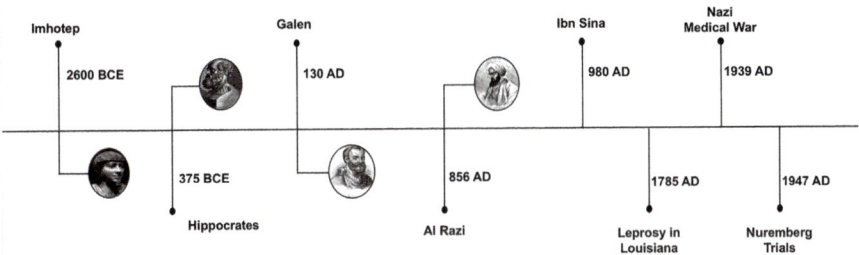

Fig. 1.2 Medical ethics history map. From Imohotep (2600 BCE) to the twentieth century

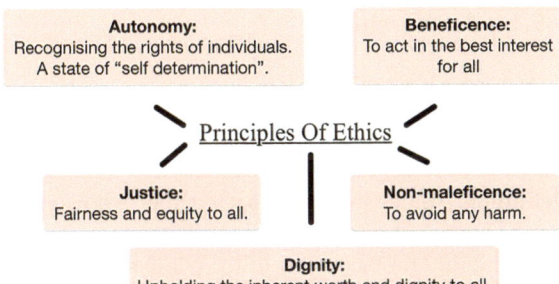

Fig. 1.3 Principles of ethics: autonomy, beneficence, justice, nonmaleficence, and dignity

One of the most influential figures in this regard was the Persian physician and philosopher Avicenna (also known as Ibn Sina) in the eleventh century. His seminal work, the Canon of medicine (law), addressed the importance of ethical considerations in health care. Avicenna emphasized the physician's duty to act with compassion, honesty, and respect for patient autonomy, echoing the Hippocratic tradition [2].

Figure 1.2 shows the medical ethics history map (from Imohotep (2600 BCE) to the twentieth century).

1.4 Principles of Ethics

Figure 1.3 presents the principles of ethics: autonomy, beneficence, justice, nonmaleficence, and dignity.

1.5 Key Ethical Considerations

Figure 1.4 *presents key ethical considerations: ethical approval, authors, plagiarism, and citations.*

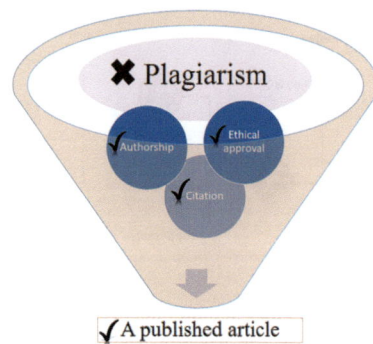

Fig. 1.4 Key ethical considerations: ethical approval, authors, plagiarism, and citations

1.5.1 Authorship

> Authorship is not a trade, it is an inspiration; authorship does not keep an office, its habitation is all out under the sky and everywhere the winds are blowing and the sun is shining and the creatures of God are free. (Mark Twain)

Definition Authorship is for each author who significantly advances science during the course of a study and contributes to it. They also share responsibility and accountability for the results of the published research.

Authorship credit carries significant academic, societal, and economic implications and is governed by regulations to ensure transparency and prevent ethical violations. Proper documentation of research findings is essential for future researchers and enhances the credibility of the work. Attention must be given to the type of study being conducted, and evaluations should be made based on the findings. The writing process can be both rewarding and demanding [3].

Key ethical concerns include:

1. **Ghost author**: Removed author to mask financial conflict of interest
2. **Guest author**: Adding a well person to increase the credibility
3. **Gift/honorary author**: Co-authorship awarded to a person who has not contributed significantly to the study
4. Sharing confidential data from peer reviews
5. Assigning the same project to multiple students to race for completion
6. Overburdening or exploiting graduate and post-doctoral students, which can harm their well-being and academic progress [3]

1.5.2 Citations [4]

Definition In research papers or speeches, quoting or paraphrasing someone else's ideas requires proper credit, helping readers locate the source [5].

1 Ethics of Research

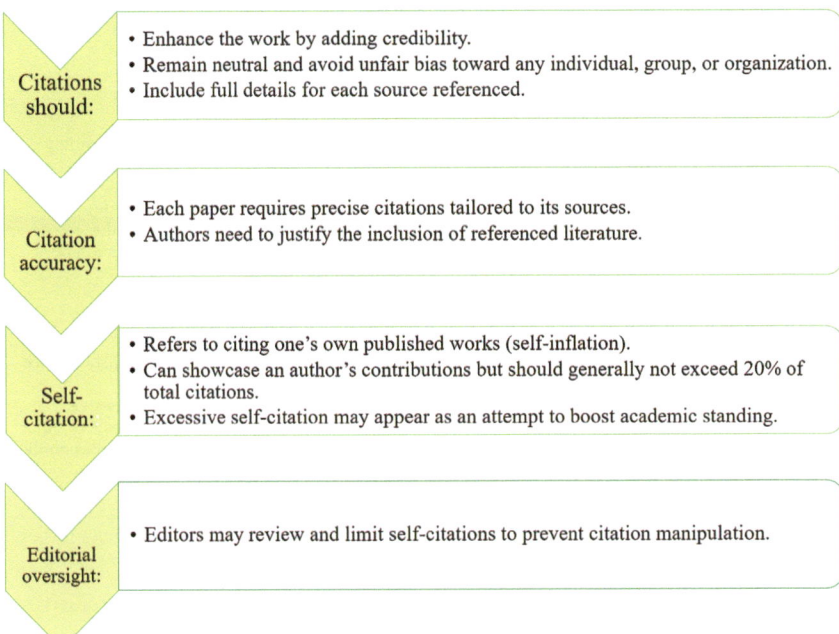

Fig. 1.5 Citations: how to properly credit someone else's work. [5]

Figure 1.5 presents citations on how to properly credit someone else's work.

1.5.3 Generative AI (Refer to Chap. 17)

The use of generative AI in research is inevitable and valuable. However, it raises many ethical concerns, particularly around confidentiality, data protection, and accountability. Generative AI models cannot take responsibility for their contributions or ensure adherence to ethical standards, making them unsuitable as substitutes for expert reviewers. Therefore, researchers must follow specific rules and ethical standards to keep the integrity of medical research.

1.5.4 Plagiarism [6]

Plagiarism is defined as the intentional or unintentional duplication of another person's words or ideas. It is unethical and may result in disciplinary actions, including expulsion.

DO'S
- Be original and not to mirror any other author's previously published work.
- When quoting someone else's remarks, it needs to be enclosed in quote marks.
- Cite the original sources of the concepts, information or techniques.
- Paraphrase when using information from other sources with proper citation.
- Get permission from all relevant professors.

DON'TS
- Edit the manuscript without informing your authors.
- Send the same work to several journals without alerting the editors to the difference or minimal modification or proper citation.
- Ask authors to cite your work.
- Duplication of an already published piece.
- Scientific papers should not involve excessive paraphrasing, synthesis of other texts, or other content from sources on the same topic.

Fig. 1.6 Plagiarism: what to do and what not to do. [7].

- **Importance of proper credit**.
  - Researchers should understand citation practices and take accurate notes.
  - Plagiarism has consequences, even if it is unintentional.
- **Consequences of plagiarism**.
  - If a document contains over **25%** plagiarized content, it may be rejected and republished with a disclaimer.
  - If plagiarism is discovered post-publication, editors may issue an "Editor's Note" or retract the piece, informing the publisher and readers.
- **Types of plagiarism**:
  - **Self-plagiarism**: Reusing one's own work without proper citation.
  - **Direct plagiarism**: Copying a passage word-for-word from another's work without acknowledgment or quotation marks.
  - **Mosaic plagiarism**: Incorporating concepts, viewpoints, or phrases from a source without credit, often with slight word changes but keeping the original structure, also known as "patch writing."
  - **Accidental plagiarism**: Failing to properly cite sources, misquoting, or paraphrasing too closely by accident. It is treated with the same severity as intentional plagiarism [7].

Figure 1.6 presents plagiarism: what to do and what not to do.
Figure 1.7 presents ways to avoid plagiarism.

- *Websites to help you in paraphrasing:* (all are free to use but some have word limit).
  https://quillbot.com/grammar-check
  https://www.paraphraser.io/

1 Ethics of Research

Fig. 1.7 Ways to avoid plagiarism [8]

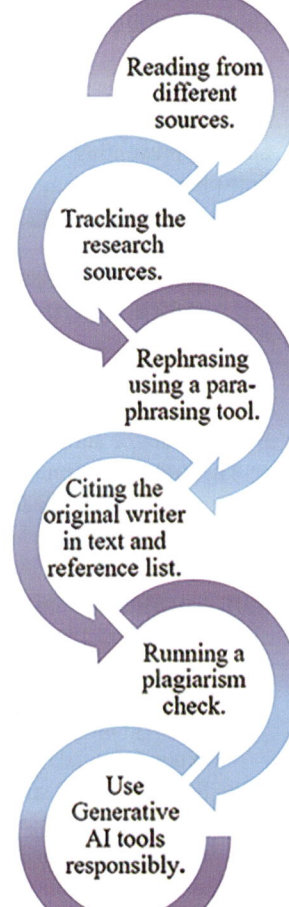

https://www.prepostseo.com/
https://rephrasetool.com/
- **Websites for detecting the percentage of plagiarism:** (all are free to use but some have word limit).
  https://www.grammarly.com/plagiarism-checker
  https://justdone.ai/try/humanize-ai
  https://www.prepostseo.com/

1.5.5 Duplicate Publications [9]

Definition Duplicate publication occurs when an author publishes a paper substantially similar to a previously published one without acknowledging the source or obtaining permission from the original copyright holder. Differences may include a new title or modified abstract, but the data and findings remain the same.

Issues
- Copyright violation: The copyright typically belongs to the journal, not the authors, preventing free republication.
- Self-plagiarism: Using material from another work without attribution [10].

1.6 Ethical Declarations That Authors Should Provide at the Journal Submission Stage [11]

- **Presubmission Considerations Related to Authorship**
  - All authors must agree on the authorship, read, and approve the manuscript.
  - The order of authors should be mutually agreed upon before submission.
  - The title page must include full names, institutional affiliations, highest degrees, and email addresses of all authors. ORCID IDs and LinkedIn profiles may also be included.
  - A corresponding author should be designated, responsible for all manuscript-related communication, and their detailed institutional affiliation (including postal address, phone number, fax number, and email) should be provided [10].
- *Other Important Declarations*
  - Authors' contribution
    - Specify individual contributions (e.g., study design, data acquisition, experiments, data analysis, and manuscript writing).
    - Follow journal-specific guidelines for declaring contributions.
  - Acknowledgments
    - Acknowledge those who provided technical help, general support, or manuscript preparation assistance.
    - If no acknowledgments are needed, state "Not applicable."
  - **Funding**
    - Declare all funding sources and their roles in the research.
    - Provide the names of funding agencies and grant numbers.
    - If no funding was received, state this.
  - **Competing Interests/Conflict of Interest**
    - Declare all financial and nonfinancial competing interests.
    - Include political, personal, religious, ideological, academic, and intellectual interests.
    - Disclose any potential conflicts of interest, including funding sources or affiliations.
    - Authors from commercial organizations should declare these interests.
  - Data Integrity
    - Ensure the accuracy and integrity of the data presented.
    - Be transparent about research methods and procedures.

- Affiliation Policy
  - All relevant affiliations (approved/supported/conducted) should be listed.
  - Moving to another affiliation (double affiliation):
    - The original one (conducted to paper) + the current one.
  - If there is no current: independent status.
- **Ethical Approval**
  - Obtain and mention ethics approval for studies involving human or animal subjects.
  - Include informed consent statements if applicable.
- Declarations Specific to Article Type:
  - Reviews: Do not require ethical approvals or informed consent. Authors should state why these are not needed for transparency.
  - Clinical Trials: Must follow CONSORT guidelines for health-related interventions. Authors should confirm adherence to these guidelines and provide the trial registry and registration number (e.g., ISRCTN).
- Statements of ethical approval for studies involving human subjects and/or animals:
  If your study involves human subjects and/or animals and also if your manuscript includes case reports/case series
  - Ethical Approval: Provide the name of the ethical approval committee/Institutional Review Board and the approval number/ID.
  - Informed Consent: State that written informed consent was obtained from study participants. If verbal consent was used, explain why written consent was not obtained.
  - Case Reports/Series: For minors, confirm written consent from legally authorized representatives/parents/guardians. If verbal consent was used, provide reasons.
  - Privacy: Do not include identifying information (e.g., images, names, initials, and hospital numbers) unless essential for scientific purposes. Obtain written consent for publication of such information. If consent is not obtained, remove personal details before submission [10].

The foundation of ethical research lies in maintaining integrity, transparency, and respect for both the scientific process and the rights of all individuals involved. As researchers, we bear the responsibility not only to uphold rigorous standards but also to foster trust within the broader community. By diligently observing ethical guidelines in authorship, citation, the responsible use of AI, and the prevention of plagiarism, we contribute to a body of work that is both credible and impactful. As science continues to evolve, so too must our commitment to ethical practices, ensuring that future generations inherit a research environment defined by respect, accountability, and a dedication to the truth.

In conclusion, Fig. 1.8 shows what should be avoided while writing a research paper.

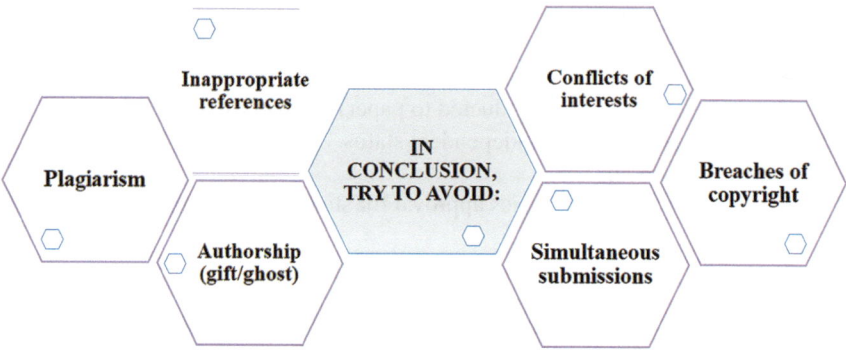

Fig. 1.8 What should be avoided while writing a research paper?

Appendix

History Map inspired by:

- Figure 2: Imhotep 2600 BCE.
  - PhilMasiello. (2023, April 15). *The Pharoah Ptahhotep: Ancient entrepreneur*. Phil Masiello Startup & Business Coach. https://philmasiello.com/the-pharoah-ptahhotep-ancient-entrepreneur/
  - University, M. B. (2023a, February 11). *The search for imhotep: Tomb of architect-turned-god remains a mystery*. ARCE. https://arce.org/resource/search-imhotep-tomb-architect-turned-god-remains-mystery/
  - Mark, J. J. (2025, March 10). *Imhotep*. World History Encyclopedia. https://www.worldhistory.org/imhotep/
  - [Imhotep—builder, physician, god]. (n.d.). https://www.researchgate.net/publication/23993511_Imhotep%2D%2Dbuilder_physician_god
- Figure 2: Hippocrates 375 BCE.
  - Hippocrates. Engraving by Peter Paul Rubens, (S.L., 1638)
  - Z;, M. (n.d.). *[Imhotep--builder, physician, god]*. Medicinski pregled. https://pubmed.ncbi.nlm.nih.gov/19203075/
  - Mark, J. J. (2025a, March 10). *Imhotep*. World History Encyclopedia. https://www.worldhistory.org/imhotep/
  - Dunn, E. (2022, July 28). *Ancient perspectives on mental illness: Insight from Hippocrates and Avicenna*. The Dunn Lab. https://www.thedunnlab.com/blog/ancient-perspectives-on-mental-illness-insight-from-hippocrates-and-avicenna
- Figure 2: Galen 130 AD.
  - BBC. (n.d.). *History - historic figures: Galen (c.130 ad - c.210 AD)*. BBC. https://www.bbc.co.uk/history/historic_figures/galen.shtml

- *Who was Galen?*. Galen Institute. (2022, January 14). https://galen.org/about/who-was-galen/#:~:text=Galen%2C%20a%20second%2Dcentury%20Greek,the%20ancient%20world%20after%20Hippocrates
- Figure 2: Al Razi 865 AD.
  - *Portrait of rhazes (Al-Razi) (AD 865–925), physician and Alchemist who lived in Baghdad*. Wellcome Collection. (n.d.-a). https://wellcomecollection.org/works/vv9w9wg4
  - Abu Bakr al-Razi (865.925.) | download scientific diagram. (n.d.-a). https://www.researchgate.net/figure/Abu-Bakr-al-Razi-865925_fig 4_318095624
  - Amr, S. S., & Tbakhi, A. (2007). *Abu Bakr Muhammad ibn Zakariya Al Razi (rhazes): Philosopher, physician and Alchemist*. Annals of Saudi medicine. https://pmc.ncbi.nlm.nih.gov/articles/PMC6074295/
  - Abu Bakr Muhammad ibn Zakariya Razi (AD 865–925) and early description of Clinical Trials - International Journal of Cardiology. (n.d.-b). https://www.internationaljournalofcardiology.com/article/S0167-5273(14)00742-6/fulltext.
  - Amin N. Daghestani, M. D., & Amin N. Daghestani, M. D. V. all articles by this author. (1997, November 1). *Al-Razi (rhazes), 865–925*. American Journal of Psychiatry. https://psychiatryonline.org/doi/10.1176/ajp.154.11.1602
- Figure 2: Ibn Sina 980 AD.
  - Amr, S. S., & Tbakhi, A. (2007b). *Ibn Sina (Avicenna): The prince of physicians*. Annals of Saudi medicine. https://pmc.ncbi.nlm.nih.gov/articles/PMC6077049/
  - Gutas, D. (2016, September 15). *Ibn Sina [avicenna]*. Stanford Encyclopedia of Philosophy. https://plato.stanford.edu/entries/ibn-sina/
- Figure 2: Leprosy in Louisiana 1785 AD.
  - Walker, Norman Purvis (1905) *An introduction to dermatology* (3rd ed.), William Wood and company Retrieved on 26 September 2010.
  - *Leprosy patients from the late 19th/20th century*. MEDizzy. (n.d.). https://medizzy.com/feed/32534066
  - BARRYLEE1. (2025, March 11). *4TeocOY*. Flickr. https://www.flickr.com/photos/21819623@N02/16031371218/
- Figure 2: Nazi medical war 1939 AD.
- *Pseudo-medical experiments in Hitler's concentration camps*. Pseudo-medical experiments in Hitler's concentration camps | Medical Review Auschwitz. (n.d.). https://www.mp.pl/auschwitz/journal/english/170062,pseudo-medical-experimens-in-hitlers-concentration-camps.
  - Mengele, Clauberg, C., Doctors, Death, Twins, Alexander, V., Guido, & Nino. (n.d.-a). *Nazi medical experiments– ppt download*. SlidePlayer. https://slideplayer.com/slide/13386594/
  - *Science and suffering: Victims and perpetrators of Nazi human experimentation*. The Wiener Holocaust Library. (2021, January 14). https://wienerholocaustlibrary.org/exhibition/science-and-suffering-victims-and-perpetrators-of-nazi-human-experimentation/

- Figure 2: Nuremburg trials 1947 AD.
  - Nazi medicine and the ethics of Human Research – The lancet. (n.d.-e). https://www.thelancet.com/journals/lancet/article/PIIS0140-6736(05)67199-1/fulltext.
  - Encyclopædia Britannica, Inc. (2025, March 14). *Nürnberg Trials*. Encyclopædia Britannica. https://www.britannica.com/event/Nurnberg-trials

References

1. Rogers, A. (2013). Research ethics. In A dictionary of human geography (pp. 348–349). Oxford University Press.
2. Young M, Wagner A. Medical Ethics. In: StatPearls [Internet]. Treasure Island (FL): StatPearls Publishing; 2024 Jan. [Updated 2024 May 7].
3. University of Leiden. History of research ethics. Available from: https://scholarlypublications.universiteitleiden.nl/access/item%3A2975092/view
4. IOP Publishing. Ethics for reviewers. Available from: https://publishingsupport.iopscience.iop.org/questions/ethics-for-reviewers/
5. Bhutta ZA. The ethics of research in developing countries. J Pak Med Assoc. 2002;52(3):105–12. PMID: 11938556; PMCID: PMC2642865.
6. Bowdoin College. Common types of plagiarism. Available from: https://www.bowdoin.edu/dean-of-students/conduct-review-board/academic-honesty-and-plagiarism/common-types-of-plagiarism.html#:~:text=Direct%20plagiarism%20is%20the%20word,for%20disciplinary%20actions%2C%20including%20expulsion
7. Mandal J, Parija SC. Ethics of authorship in scientific publications. Trop Parasitol. 2013;3(2):104–5. https://doi.org/10.4103/2229-5070.122108. PMID: 24470992; PMCID: PMC3889085.
8. National Institute of Environmental Health Sciences. Ethical issues in research and publication. Available from: https://www.niehs.nih.gov/research/resources/bioethics/whatis
9. City, University of London. Research ethics approval guidelines. Available from: https://info.lse.ac.uk/staff/divisions/research-and-innovation/research/research-ethics/Research-Ethics-Submission-System
10. Castleton University. Information ethics: citing sources and fair use. Castleton University Library. [Updated 2024 Sep 13; cited 2024 Sept 13]. Available from: https://www.castleton.edu/library/information-literacy-graduation-standard/information-literacy-tutorial/information-ethics-citing-sources-and-fair-use/#:~:text=When%20you%20quote%20or%20paraphrase,locate%20the%20source%20you%20cited
11. University of Leeds. Ethical approval requirements for research involving humans or animals. Available from: https://secretariat.leeds.ac.uk/research-ethics/how-to-apply-for-research-ethics-approval/

How to Find a Meaningful Research Question

2

Neveen A. Kohaf, Esraa Y. Salem, Islam Mohammad Shehata, Omar Viswanath, and Alan David Kaye

> *In the realm of knowledge, every question plants the seed of research; choose yours wisely.*

Abbreviations

AI   Artificial Intelligence
PICO   Population, Intervention, Comparison, Outcome

N. A. Kohaf (✉)
Clinical Pharmacy Department, Faculty of Pharmacy (Girls), Al-Azhar University, Cairo, Egypt

Clinical Pharmacy Department, Faculty of Pharmacy, Al Baha University,
Al Baha, Saudi Arabia
e-mail: nevenabdo@azhar.edu.eg

E. Y. Salem
Faculty of Medicine, Modern University for Technology and Information, Cairo, Egypt

I. M. Shehata
Department of Anesthesiology, Faculty of Medicine, Ain Shams University Cairo,
Cairo, Egypt
e-mail: islam.shehata@med.asu.edu.eg

O. Viswanath
Department of Anesthesiology, Creighton University School of Medicine, Phoenix, AZ, USA

Mountain View Headache and Spine Institute, Phoenix, AZ, USA

A. D. Kaye
Department of Anesthesiology, Louisiana State University Health Sciences Center
at Shreveport, Shreveport, LA, USA

Department of Pharmacology, Toxicology, and Neurosciences, Louisiana State
University Health Sciences Center at Shreveport, Shreveport, LA, USA
e-mail: alan.kaye@lsuhs.edu

© The Author(s), under exclusive license to Springer Nature Switzerland AG 2025
I. M. Shehata, O. Viswanath (eds.), *How to Successfully Publish a Manuscript*,
https://doi.org/10.1007/978-3-031-92538-2_2

FINERMAPS	Feasible, Interesting, Novel, Ethical, Relevant, Manageable, Appropriate, Potential Value, Publishable, Systematic
PubMed	Public/Publisher Medline
PROSPERO	International Prospective Register of Systematic Reviews
INPLASY	International Platform of Registered Systematic Review and Meta-analysis Protocols

2.1 Introduction

A hypothesis in research methodology is a tentative, testable explanation or prediction about the relationship between variables in a study. It serves as a guiding statement for the research, proposing a potential outcome based on existing knowledge or theories. This educated conjecture directs the investigation and can be either supported or refuted through systematic data collection and analysis. Hypotheses play a crucial role in shaping the research design, determining appropriate methods, and providing a framework for interpreting results [1, 2].

2.1.1 Characteristics of a Good Hypothesis

- *Testability*: An effective hypothesis necessitates validation through empirical data collection and analysis.
- *Variables*: It distinctly defines the independent and dependent variables pertinent to the study.
- *Clarity*: The hypothesis must be articulated in a concise and understandable language, eliminating ambiguity.

2.2 Definition

The research question is the foundation upon which the entire study is built. It determines the scope and focus of the investigation, shapes the methodology, and ultimately influences the significance and contribution of the findings [3].

2.3 Importance

Formulating an effective research question is a crucial first step of any research endeavor. A well-crafted research question not only provides clear direction for the study but also serves as a guidepost throughout the research process (Fig. 2.1) [4].

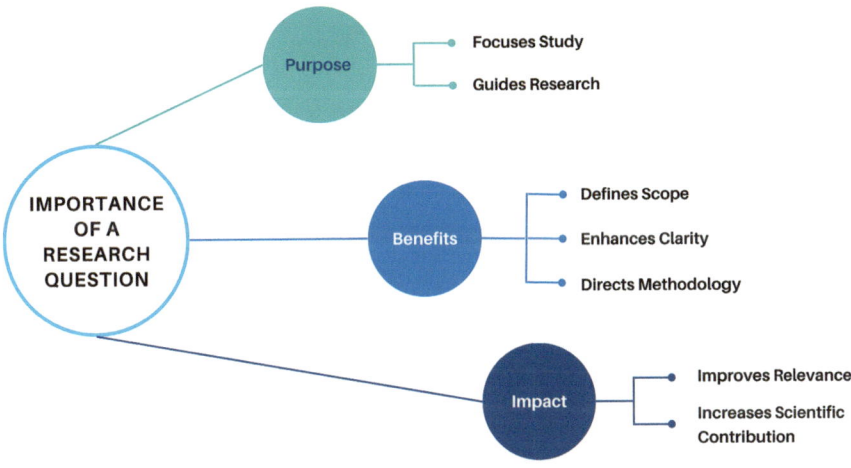

Fig. 2.1 Importance of research questions

2.4 Characteristics of Research Questions

Identifying a meaningful and impactful research question can be a challenging task, as it requires [5]:

1. A well-rooted understanding of the existing literature
2. An awareness of current research trends
3. The ability to identify knowledge gaps

Furthermore, an effective research question should be precise and complex enough to warrant in-depth analysis, rather than a simple "yes" or "no" answer. It should also be arguable or testable, with the potential for multiple interpretations and the ability to withstand scrutiny [6]. This ensures that the research question is not merely a statement of fact but a genuine inquiry that can be explored through rigorous investigation (Fig. 2.2) [7].

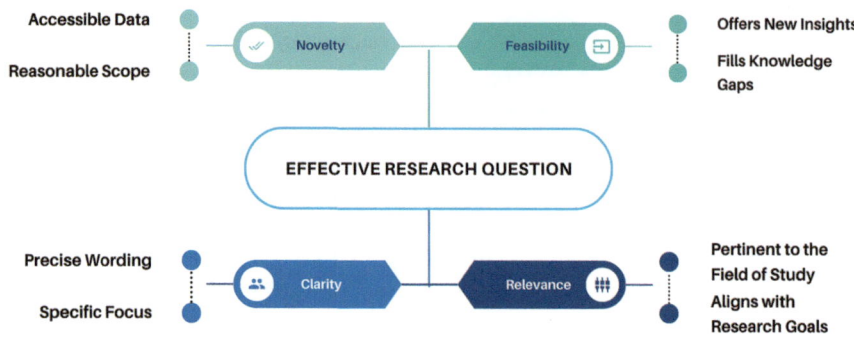

Fig. 2.2 Characteristics of effective research questions

2.5 Strategies for Identifying Research Questions

Identifying a meaningful research question often requires a multifaceted approach. Here are three strategies you can follow:

1. Conducting a **thorough review** of the existing literature:

    This involves not only reading the relevant studies but also analyzing the gaps, limitations, and unanswered questions within the field. By identifying these knowledge gaps, researchers can develop specific research questions that have the potential to make a significant contribution to the existing body of knowledge.

    Example: meta-analysis/systematic review [8, 9].

2. **Up to date** with the latest trends and developments in the field:

    This can be achieved by regularly reading academic journals, attending conferences, and engaging with the research community. Being informed of the current state of the field allows researchers to identify the emerging areas of interest and formulate research questions that address pressing issues or unexplored topics.

    Examples: Techniques: hypotensive predictive index [10], COVID: proceed or postpone [11].

3. Inspiration from **your own observations**, experiences, and intuitions:

    By reflecting on your own interests, curiosities, and personal experiences, you can identify research questions that resonate with yourself and have the potential to generate meaningful insights. This approach can lead to the development of unique and innovative research questions that challenge existing assumptions and push the boundaries of the field.

    Examples: Case report: pneumothorax [12].

2.6 Stepwise Approach to Developing an Effective Research Question

The stepwise approach to developing an effective research question involves a systematic process of **narrowing** down a broad topic of interest to a specific and focused inquiry. The next crucial step is converting the focused topic into a **question form**, ensuring it is researchable, feasible, specific, complex, and relevant to the field of study.

- **The question is then evaluated** using criteria such as the **FINERMAPS** framework to assess its effectiveness:
  - **F**easible
  - **I**nteresting
  - **N**ovel
  - **E**thical
  - **R**elevant
  - **M**anageable
  - **A**ppropriate
  - **P**otential value
  - **P**ublishability
  - **S**ystematic

The process **concludes** with:

1. Refining and rewriting the question as needed.
2. Breaking it down into key concepts.
3. Specifying the population, intervention/exposure, and outcomes of interest.

This approach helps researchers develop a clear, focused, and effective research question that can guide their study and contribute meaningfully to their field of inquiry.

- **Now you have the final product—what is the best way to assess?** [13]
  - *Conduct Expert Reviews*
    - Share the questions with peers and mentors in your field.
    - Ask them to assess the questions based on the defined criteria.
    - Solicit feedback on clarity, significance, and methodological approach.
    - Consider organizing a formal expert panel review session.
  - *Use Assessment Rubrics*
    - Create a scoring system for each criterion (e.g., 1–5 scale).
    - Rate questions on factors like clarity, feasibility, originality, etc.
    - Calculate overall scores to compare and rank different questions.
    - Identify areas for improvement based on low-scoring criteria.
  - *Perform Literature Reviews*
    - Search academic databases and Google Scholar for related work.

- Assess how the questions align with or expand on existing research.
- Identify any gaps in the literature the questions could address.
- Determine if similar questions have already been thoroughly answered.
– *Solicit Peer Feedback*
  - Engage colleagues in assessing your research questions.
  - Present questions at departmental seminars or conferences.
  - Discuss with peers in your research group or lab meetings.
  - Share on academic social networks like ResearchGate for input.
  - Consider publishing questions as a preprint to gather community feedback.

2.7 Searching Databases

A database is a collection of journals. One way of finding a research idea is searching a specific database. According to the subject of your search, you can find databases that help your search best, for instance, PubMed, Scopus, Google Sholar, and Cochrane Library are well-known widely used health science databases, while in engineering, IEEE Xplore, Scopus, Google Sholar, and Web of Science are usually used. You can always find the right database(s) if you search Google [14]. To start searching databases, you need to have at least one item of the PICO, which stands for population, intervention, comparison, and outcome.

So, let us say you have a specific population in mind, for instance, **X**:

Step 1: You start by opening suitable databases and searching the title and abstracts by typing **X** in the search bar and adding [Title/Abstract] or its equivalent in each database.
Step 2: Order the results with the newest first.
Step 3: Use the study design filters to display the best-fitting studies for your review type. For instance, randomized clinical trials filter can be used if you are aiming to do a systemic review of randomized clinical trials.
Format Example:
PubMed: (**X** [Title/Abstract])
Scopus: TITLE-ABS ("**X**")
Web of Science: TS = (**X**)
Google Scholar: **X**
Cochrane Library: ("**X**"):ti,ab,kw (tiTLE, abSTRACT, kEY wOrds).
Step 4: Recognize patterns by going through the articles and writing down repeated terms or ideas.
Step 5: After collecting enough ideas, you can start searching by typing **X** AND one of the ideas you found, for instance, vitamin D deficiency, each followed by [Title/Abstract] or its equivalent in each database. Here, the connector AND is one of the Boolean operators; it retrieves articles containing both the first and the second terms.
Format Example:
PubMed: (**X** [Title/Abstract]) AND (vitamin D[Title/Abstract])

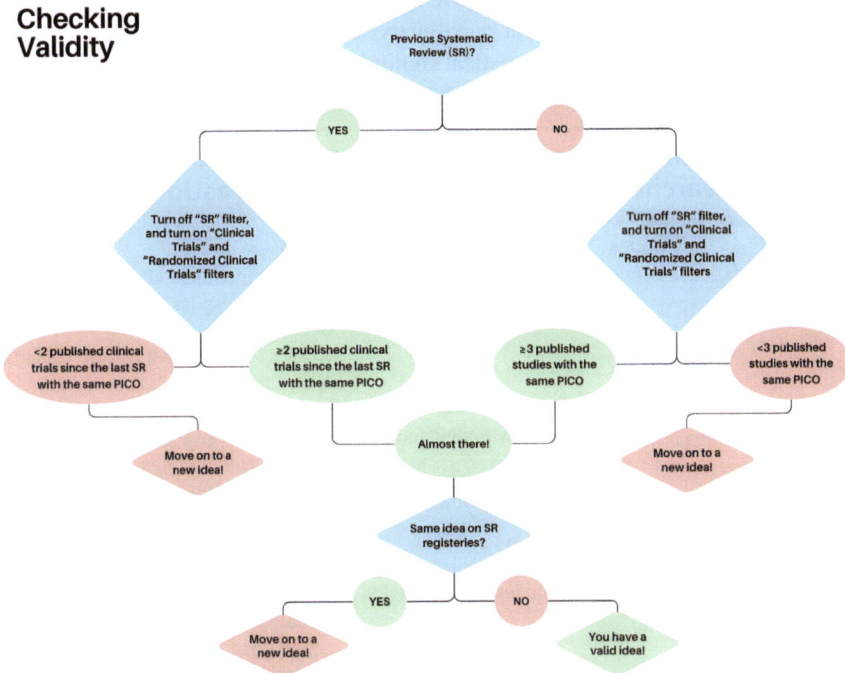

Fig. 2.3 Checking the validity of a review idea

Scopus: (TITLE-ABS ("**X**") AND TITLE-ABS ("vitamin D"))
Web of Science: TS = (**X**) AND TS = (vitamin D)
Google Scholar: **X** AND vitamin D
Cochrane Library: ("**X**"):ti,ab,kw AND ("vitamin D"):ti,ab,kw

N.B.: The "Advanced Search" option is provided by most databases to aid in writing these formats, so you would rather click buttons rather than manually write the fields.

Step 6: Filter the retrieved articles by the systematic reviews filter to make sure there are enough studies for the validity of the idea.

Step 7: Check the validity of your idea by making sure no recent reviews of the same PICO have been recorded either on databases or systematic review registries such as PROSPERO, INPLASY, and Open Science Framework Registries and Protocols.io [15]. Now, you may encounter multiple scenarios while validating your idea; follow the flowchart in (Fig. 2.3) to find your way out.

There are no specific guidelines on the minimum number of studies needed to write a systematic review; however, some scientists suggest having at least three studies with the same PICO, while others deny a minimum number of studies and encourage researchers to report the systematic process even if they found no studies as this will encourage researchers to give those topics more attention, and these are called empty reviews [16]; as for meta-analyses, scholars suggest having a minimum of five to ten studies for statistical significance [17, 18].

2.8 Artificial Intelligence (AI) Offers Several Tools to Assist Researchers in Formulating Research Questions

Tool	Link	Character
Research question generators	(https://www.writecream.com/research-question-generator)	Crafting focused and relevant research questions based on user input Allows users to input research details and generates tailored questions instantly
AI research assistants	https://elicit.com/?redirected=true	Retrieving relevant papers and summarizing findings
Prompt generators	(https://www.taskade.com/generate/ai-prompts/research-question-prompt)	Provide prompts to stimulate the development of research questions, enhancing creativity and depth in inquiry
InfraNodus extension for knowledge graphs	https://infranodus.com/extension	Visualizes text as a network, revealing structural gaps and generating research questions using built-in AI models
Musely's research question generator	https://musely.ai/tools/research-question-generator	Generates a list of potential research questions, aiding in the refinement of research objectives
Hyperwrite's AI research question formulator	https://www.hyperwriteai.com/aitools/ai-research-question-formulator	Create precise and relevant research questions based on your area of study and initial topic ideas

2.9 Conclusion

Identifying a meaningful research question is a critical first step in any research endeavor. It is the foundation upon which the rest of the research is based. A well-crafted research question provides clear direction, shapes the research methodology, and ultimately determines the significance and contribution of the study. By employing strategies such as literature review, staying informed about current trends, and drawing on personal experiences, researchers can develop research questions that are clear, focused, and have the potential to advance the field of study. Additionally, evaluating the research question against established criteria can help ensure its feasibility, relevance, and impact. By mastering the art of research question formulation, researchers can lay the foundation for successful and impactful studies.

References

1. Kumar R. Research methodology: a step-by-step guide for beginners. London: Sage; 2005.
2. Bryman A. Social research methods. 5th ed. Oxford: Oxford University Press; 2016.
3. Singh S. How to identify research questions for your study. 2021. Available from: https://researcher.life/blog/article/tips-identify-meaningful-research-question/.
4. Wordvice. How to write a good research question (w/examples) 2024. Available from: https://blog.wordvice.com/how-to-write-a-hypothesis-or-research-question/.
5. IUL. Narrowing a topic and developing a research question. n.d. Available from: https://libraries.indiana.edu/sites/default/files/Develop_a_Research_Question.pdf.
6. Insights E. In which section of a paper should the research question be included? 27 Dec 2017. Available from: https://www.editage.com/insights/in-which-section-of-the-paper-should-the-research-question-be-included.
7. MUL. Developing research questions. n.d. Available from: https://www.monash.edu/library/help/assignments-research/developing-research-questions.
8. Elsaeidy AS, Ahmad AHM, Kohaf NA, et al. Efficacy and safety of ketamine-dexmedetomidine versus ketamine-propofol combination for periprocedural sedation: a systematic review and meta-analysis. Curr Pain Headache Rep. 2024;28:211–27. https://doi.org/10.1007/s11916-023-01208-0.
9. Shehata IM, Kohaf NA, ElSayed MW, et al. Ketamine: pro or antiepileptic agent? A systematic review. Heliyon. 2024;10 https://doi.org/10.1016/j.heliyon.2024.e24433.
10. Shehata IM, Alcodray G, Essandoh M, et al. Con: routine use of the hypotension prediction index in cardiac, thoracic, and vascular surgery. J Cardiothorac Vasc Anesth. 2021;35:1237–40. https://doi.org/10.1053/j.jvca.2020.09.128.
11. Shehata IM, Elhassan A, Jung JW, et al. Elective cardiac surgery during the COVID-19 pandemic: proceed or postpone? Best Pract Res Clin Anaesthesiol. 2020;34:643–50. https://doi.org/10.1016/j.bpa.2020.07.005.
12. Shehata IM, Hashim RM. Pneumomediastinum, pneumothorax, pneumoperitoneum, and subcutaneous emphysema complicating extubation of a difficult airway using an airway exchange catheter: is oxygen insufflation innocent?: a case report. A&A Pract. 2020;14:e01228. https://doi.org/10.1213/xaa.0000000000001228.
13. Hulley SB, Cummings SR, Browner WS, Grady DG, Newman TB. Designing clinical research. Philadelphia: Lippincott Williams & Wilkins; 2013.
14. Welch medical library guides: expert searching: which databases to use. 2024. Available from: https://browse.welch.jhmi.edu/searching/databases-by-subject.
15. Pieper D, Rombey T. Where to prospectively register a systematic review. Syst Rev. 2022;11:8. https://doi.org/10.1186/s13643-021-01877-1.
16. Gray R. Empty systematic reviews: identifying gaps in knowledge or a waste of time and effort? Nurse Author Ed. 2021;31:42–4. https://doi.org/10.1111/nae2.23.
17. Myung SK. How to review and assess a systematic review and meta-analysis article: a methodological study (secondary publication). J Educ Eval Health Prof. 2023;20:24. https://doi.org/10.3352/jeehp.2023.20.24.
18. Huessin H. What is the minimum number of studies to be included in a systematic review? 2015. Available from: https://www.researchgate.net/post/What-is-the-minimum-number-of-studies-to-be-included-in-a-systematic-review.

Title and Keywords

3

Alaa Mohamed Ibrahim, Islam Mohammad Shehata, Omar Viswanath, and Latha Ganti

3.1 Title

Don't judge a book by its cover. (George Eliot)

However, your paper is going to be judged by your title; therefore, the title is the single most important line of your publication [1].

3.1.1 Importance

For every 500 people who just read the title, only one person reads the whole paper, which indicates that the majority of papers are read by their titles alone [2].

1. It is the first thing the reader is going to approach to assess the paper as a whole.
2. It describes the content of the paper.
3. It helps with making the researcher choose your paper.

A. M. Ibrahim (✉)
Faculty of Medicine, Modern University for Technology and Information, Cairo, Egypt
e-mail: alaa.94926@Medicine.mti.edu.eg

I. M. Shehata
Department of Anesthesiology, Faculty of Medicine, Ain Shams University Cairo, Cairo, Egypt
e-mail: islam.shehata@med.asu.edu.eg

O. Viswanath
Department of Anesthesiology, Creighton University School of Medicine, Phoenix, AZ, USA

Mountain View Headache and Spine Institute, Phoenix, AZ, USA

L. Ganti
Research Orlando College of Osteopathic Medicine, Winter Garden, FL, USA

© The Author(s), under exclusive license to Springer Nature Switzerland AG 2025
I. M. Shehata, O. Viswanath (eds.), *How to Successfully Publish a Manuscript*, https://doi.org/10.1007/978-3-031-92538-2_3

Types of titles [2]

	Descriptive (neutral)	Declarative	Compound (hanging title)	Interrogative
Definition	It includes the elements of the research work	It includes the main findings or the core message of the study	It contains the main title and a subtitle that are separated by a colon (:)	It is in the form of a question
Characters	It outlines the topic of the paper but does not reveal the main findings	It has a psychological effect which gives the impression that the conclusions of the study are generally valid and applicable. It helps authors in selecting a more suitable paper while conducting their research **N.B.** Some journals do not accept declarative titles (e.g., New England *Journal of Medicine*)	It is more useful for complex studies as it provides additional relevant information	It has a psychological factor which makes the title more attractive as the reader becomes more curious to know the answer to that question
Examples	Airway management considerations in patients with vocal cord implants [3]	Intraoperative hypotension increased risk in the oncological patients [4]	Left atrial appendage occlusion: transesophageal echocardiography versus intracardiac echocardiography-pro: intracardiac echocardiography [5]	Elective cardiac surgery during the COVID-19 pandemic: proceed or postpone? [6]

You can always mix between the different types of titles to make your title more interesting.

3.1.2 Formatting of the Title

- Follow the author guidelines of your journal.
- The title should be a single, standalone sentence which can be separated into two parts with a colon or hyphen.
- The typical number of words can be up to 20 words and vary according to the guidelines of the journal that you are choosing.

3.1.3 What to Do?

1. Attract the reader's attention while piquing their curiosity (you can do this by choosing an **interrogative title**).
2. **Read a lot,** as reading is the finest teacher of writing and do not get frustrated when it requires a lot of time as writing the title is not an easy process.
3. **Highlight your novel idea:**
   This gives your paper more chances to be read as it would be a new area of research.
4. **Choose your words wisely:**
   Include relevant words in your title that reflect the content of your article. This will help with the search engine and will give your paper more recognition.
5. **Be short yet coherent:**
   A 2010 study conducted regarding the length of titles showed a high concentration of papers with short titles and many citations, as well as a high concentration of papers with long titles and few citations. For the top 20,000 most highly cited papers published in 2010, papers with shorter titles received more citations; so, although longer titles may provide more information regarding the content, they reduce the interest generated [7].
6. **Be specific:**
   - Identify the focus of your research.
   - Highlight the key points to attract the right audience.
7. **Be simple yet scientific:**
   - Stay away from difficult, overcomplicated terms because English is not necessarily the first language for all readers.

3.1.4 What to Avoid?

1. **Avoid clickbait titles:**
   Clickbait definition:
   Misleading content designed to attract attention and encourage users to click on a link, and it may not accurately represent the actual content.
   - Readers frequently perceive the usage of clickbait as a deceptive strategy employed by publishers.
   - It causes the reader to reject it and consider it manipulation.
   - It will affect your credibility [8].
2. **Avoid abbreviations:**
   Abbreviations should be avoided in titles unless they are very common and do not need defining.
3. **Avoid being repetitive:**
   Choosing a unique title will make your paper more distinguishable.

3.1.5 Running Title

Definition
'The title or abbreviated title of a volume printed at the top or the bottom pages or sometimes of all text pages according to the journal guidelines.'

Be Aware
- It mostly cannot be longer than 50 characters (including spaces), and it varies from one journal to another [2].
- Journals might require a running title from you before publication as it helps the reader to know more information about your paper by simply looking at the top/bottom of the page they are in.

3.1.6 You Can Use AI Tools to Help You with Formulating Your Title

- You can ask any LLM such as chat GPT to recommend a title for your paper, just upload your paper or its introduction, and ask for title recommendations. You can even ask for specific criteria in the title.
- Here are some tools that can help you in creating your title:
  - AvidnoteAI
    https://avidnote.com/
  - EssayGPT
    https://essaygpt.hix.ai/tools/research-title-generator?utm_source=chatgpt.com
  - GalaxyAI
    https://galaxy.ai/ai-research-title-generator?utm_source=chatgpt.com
  - X.writefull
    https://x.writefull.com/title-generator?utm_source=chatgpt.com
  - HyperWrite
    https://www.hyperwriteai.com/aitools/academic-paper-title-generator?utm_source=chatgpt.com

Some of these websites give you options for choosing the target audience, the tone of voice whether you want it to be professional, engaging, formal, or academic. This will help you in customizing your title.

3.2 Keywords

Definition
Keywords are terms that are relevant to your paper which help in finding the paper in the database and search engines.

Importance
- Keywords impact how many researchers find your paper.
- They facilitate the indexing process in academic journals.

How to Choose Keywords
1. Follow the author guidelines of your journal regarding the number and style.
2. Some journals might provide you with a list of keywords to choose from.
3. Use MeSH:
    MeSH (medical subject headings) is a tool that will help you find MeSH terms relevant to your text.

To Learn How to Use MeSH on Demand Tool
- You can visit the following link to use MeSH on demand tool; it will provide you with detailed steps:
  https://www.nlm.nih.gov/pubs/techbull/mj14/mj14_mesh_on_demand.html
- You can visit the following link; it will provide you with a detailed video explanation on how to use this tool:
  https://www.nlm.nih.gov/oet/ed/mesh/meshondemand.html

It is recommended to use MeSH words that are not already in your title, as including it does not increase the findability of your paper because words that are present in your title will be already in the search engine.

If you encounter a difficulty in picking your keywords, here are some examples which might help you:

- Ask yourself what keywords you would type in the search engine if you were searching for your exact topic.
- You can use the ideas given in Fig. 3.1 according to the subject of your paper.

Here are some examples:

1. **Title**: Evolving Concepts of Pain Management in Elderly Patients [9]
    Keywords: Acute pain, Chronic pain, Elderly pain, Geriatric pain, Pain management
    - Types of the subject (pain): Acute pain; Chronic pain
    - The population targeted: Elderly pain; Geriatric pain
    - The target: Pain management
2. **Title:** Ketamine: Pro or Antiepileptic Agent? A Systematic Review [10]
    Keywords: Ketamine, Epilepsy, Anesthesia, Sedation
    - The drug used: Ketamine
    - The disease: Epilepsy
    - The physician specialty: Anesthesia
    - The procedure (target of the study): Sedation

Fig. 3.1 Keyword ideas

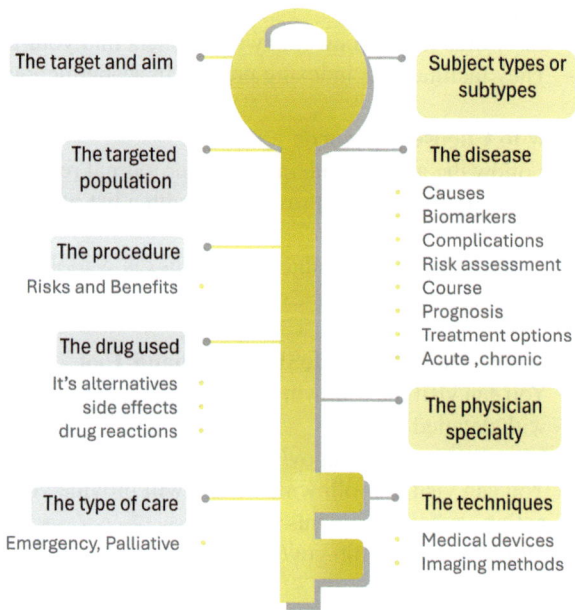

3. **Title:** Mitral Regurgitation in Patients Undergoing Noncardiac Surgery [11]

   Keywords: Mitral valve, Degenerative mitral regurgitation, Ischemic mitral regurgitation, Noncardiac surgery, **Heart failure, Intraoperative assessment, Cardiovascular risk**

   Note: Here, the writer has used words that are not included in the title which we recommended before, such as mentioning the risk assessment and complications.
   - Complication: Heart failure
   - Risk assessment: Intraoperative assessment, Cardiovascular risk
     Also, the author included:
   - The focus of the study: Mitral valve
   - The problem targeted: Degenerative mitral regurgitation, Ischemic mitral regurgitation
   - The scope of the surgery: Noncardiac surgery

4. **Title:** Airway Management Considerations in Patients with Vocal Fold Implants [3]

   Keywords: Vocal fold implants, Airway management, Anesthesia, Endotracheal intubation, Ultrasound
   - The target of the study: Airway management
   - The physician specialty: Anesthesia
   - The scope of the study: Vocal fold implants
   - The procedures performed: Endotracheal intubation
   - The method of imaging used: Ultrasound

References

1. Langford CA, Pearce PF. Increasing visibility for your work: the importance of a well-written title. J Am Assoc Nurse Pract. 2019;31(4):217–8.
2. Bahadoran Z, Mirmiran P, Kashfi K, Ghasemi A. The principles of biomedical scientific writing: title. Int J Endocrinol Metab. 2019;17(4):e98326.
3. Shehata IM, Masood W, Daebis A, Gamal I, Urits I, Viswanath O, et al. Airway management considerations in patients with vocal fold implants. Front Anesthesiol [Internet]. 2023 [cited 2024 Nov 8];2. Available from: https://www.frontiersin.org/journals/anesthesiology/articles/10.3389/fanes.2023.1209229/full
4. Shehata IM, Elhassan A, Munoz DA, Okereke B, Cornett EM, Varrassi G, et al. Intraoperative hypotension increased risk in the oncological patient. Anesthesiol Pain Med. 2021;11(1):e112830.
5. Shehata I, Essandoh M, Hummel J, Amer N, Saklayen S. Left atrial appendage occlusion: transesophageal echocardiography versus intracardiac echocardiography—pro: intracardiac echocardiography. J Cardiothorac Vasc Anesth. 2024;38(1):316–9.
6. Shehata IM, Elhassan A, Jung JW, Urits I, Viswanath O, Kaye AD. Elective cardiac surgery during the COVID-19 pandemic: proceed or postpone? Best Pract Res Clin Anaesthesiol. 2020;34(3):643–50.
7. Letchford A, Moat HS, Preis T. The advantage of short paper titles. R Soc Open Sci. 2015;2(8):150266.
8. Mukherjee P, Dutta S, De Bruyn A. Did clickbait crack the code on virality? J Acad Mark Sci. 2022;50(3):482–502.
9. Kaye AD, Kweon J, Hashim A, Elwaraky MM, Shehata IM, Luther PM, et al. Evolving concepts of pain management in elderly patients. Curr Pain Headache Rep. 2024;28(10):999–1005.
10. Shehata IM, Kohaf NA, ElSayed MW, Latifi K, Aboutaleb AM, Kaye AD. Ketamine: pro or antiepileptic agent? A systematic review. Heliyon. 2024;10(2):e24433.
11. Richter EW, Shehata IM, Elsayed-Awad HM, Klopman MA, Bhandary SP. Mitral regurgitation in patients undergoing noncardiac surgery. Semin Cardiothorac Vasc Anesth. 2022;26(1):54–67.

Introduction

4

Esraa Elbanna, Marina Ramzy, Islam Mohammad Shehata,
Omar Viswanath, and Farnad Imani

Well-begun is half done

—Aristotle

4.1 Definition

Introduction serves as the opening section of your research paper where you set up the topic and establish the context for the audience. It could also be referred to as the formal presentation of the topic [1].

E. Elbanna (✉)
Alexandria Main University Hospital, Alexandria, Egypt
e-mail: Esraa.awad1801@alexmed.edu.eg

M. Ramzy
Faculty of Medicine, Alexandria University, Alexandria, Egypt

I. M. Shehata
Department of Anesthesiology, Faculty of Medicine, Ain Shams University Cairo, Cairo, Egypt
e-mail: islam.shehata@med.asu.edu.eg

O. Viswanath
Department of Anesthesiology, Creighton University School of Medicine, Phoenix, AZ, USA

Mountain View Headache and Spine Institute, Phoenix, AZ, USA

F. Imani
MD FIPP EDPM Pain Research Center, Department of Anesthesiology and Pain Medicine, Iran University of Medical Sciences, Tehran, Iran

© The Author(s), under exclusive license to Springer Nature Switzerland AG 2025
I. M. Shehata, O. Viswanath (eds.), *How to Successfully Publish a Manuscript*,
https://doi.org/10.1007/978-3-031-92538-2_4

4.2 The Importance [2]

The importance is clearly presented in Fig. 4.1.

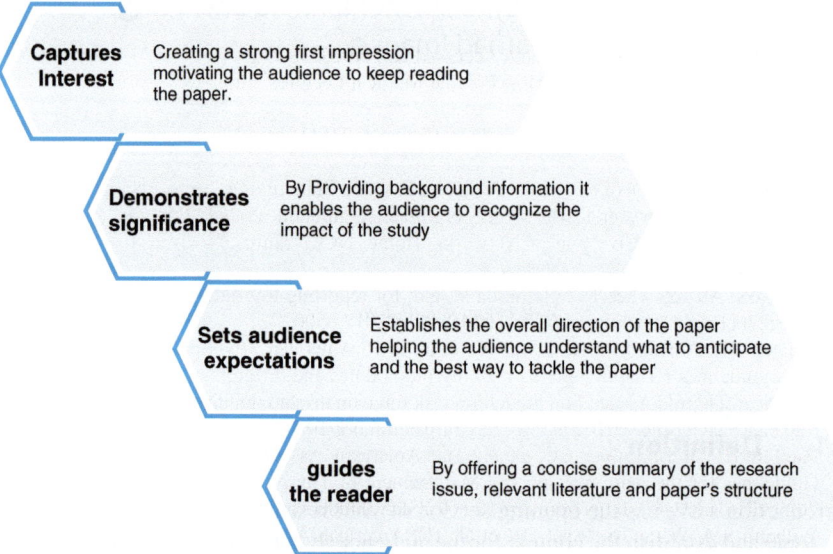

Fig. 4.1 The importance of introduction

4.3 The Structure [3]

The structure is illustrated in Fig. 4.2.
 Here is a real example [4]
 As shown in Fig. 4.3.

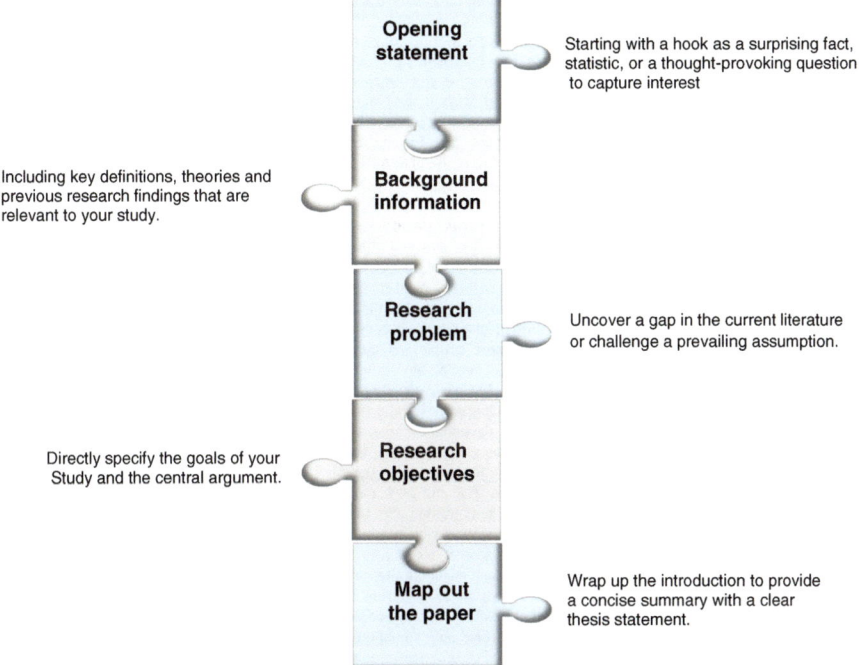

Fig. 4.2 The structure of introduction

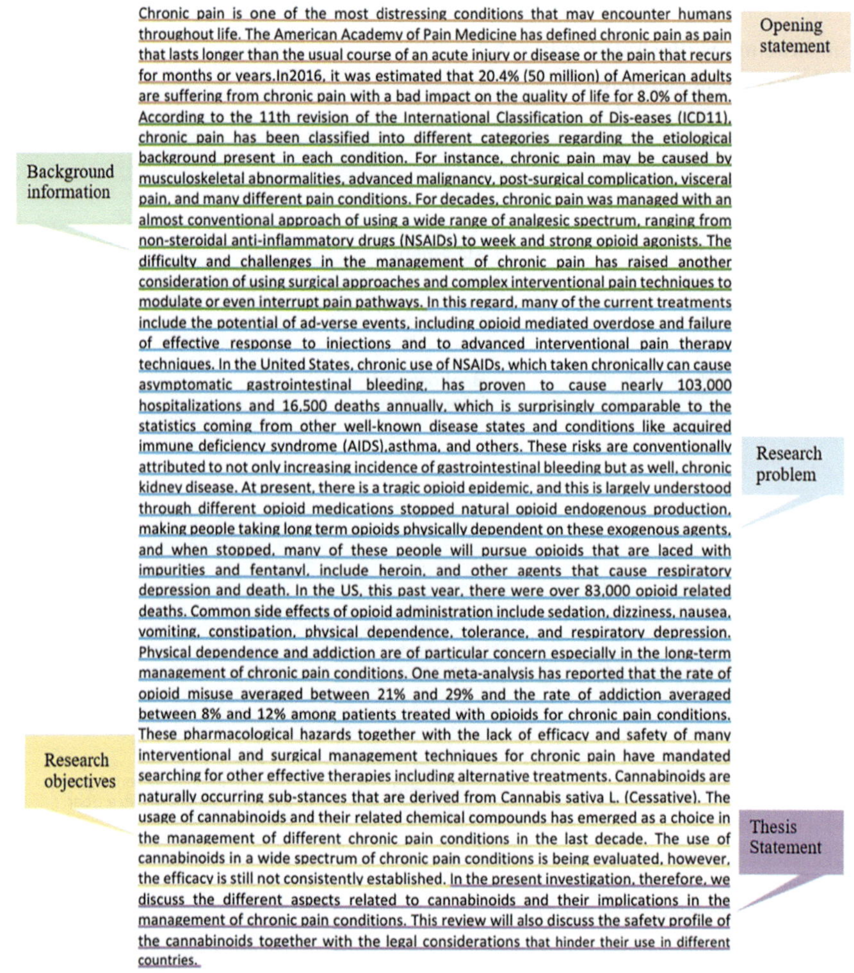

[Opening statement] Chronic pain is one of the most distressing conditions that may encounter humans throughout life. The American Academy of Pain Medicine has defined chronic pain as pain that lasts longer than the usual course of an acute injury or disease or the pain that recurs for months or years. In 2016, it was estimated that 20.4% (50 million) of American adults are suffering from chronic pain with a bad impact on the quality of life for 8.0% of them.

[Background information] According to the 11th revision of the International Classification of Dis-eases (ICD11), chronic pain has been classified into different categories regarding the etiological background present in each condition. For instance, chronic pain may be caused by musculoskeletal abnormalities, advanced malignancy, post-surgical complication, visceral pain, and many different pain conditions. For decades, chronic pain was managed with an almost conventional approach of using a wide range of analgesic spectrum, ranging from non-steroidal anti-inflammatory drugs (NSAIDs) to week and strong opioid agonists. The difficulty and challenges in the management of chronic pain has raised another consideration of using surgical approaches and complex interventional pain techniques to modulate or even interrupt pain pathways. In this regard, many of the current treatments include the potential of ad-verse events, including opioid mediated overdose and failure of effective response to injections and to advanced interventional pain therapy techniques. In the United States, chronic use of NSAIDs, which taken chronically can cause asymptomatic gastrointestinal bleeding, has proven to cause nearly 103,000 hospitalizations and 16,500 deaths annually, which is surprisingly comparable to the statistics coming from other well-known disease states and conditions like acquired immune deficiency syndrome (AIDS), asthma, and others.

[Research problem] These risks are conventionally attributed to not only increasing incidence of gastrointestinal bleeding but as well, chronic kidney disease. At present, there is a tragic opioid epidemic, and this is largely understood through different opioid medications stopped natural opioid endogenous production, making people taking long term opioids physically dependent on these exogenous agents, and when stopped, many of these people will pursue opioids that are laced with impurities and fentanyl, include heroin, and other agents that cause respiratory depression and death. In the US, this past year, there were over 83,000 opioid related deaths. Common side effects of opioid administration include sedation, dizziness, nausea, vomiting, constipation, physical dependence, tolerance, and respiratory depression. Physical dependence and addiction are of particular concern especially in the long-term management of chronic pain conditions. One meta-analysis has reported that the rate of opioid misuse averaged between 21% and 29% and the rate of addiction averaged between 8% and 12% among patients treated with opioids for chronic pain conditions.

[Research objectives] These pharmacological hazards together with the lack of efficacy and safety of many interventional and surgical management techniques for chronic pain have mandated searching for other effective therapies including alternative treatments. Cannabinoids are naturally occurring sub-stances that are derived from Cannabis sativa L. (Cessative). The usage of cannabinoids and their related chemical compounds has emerged as a choice in the management of different chronic pain conditions in the last decade. The use of cannabinoids in a wide spectrum of chronic pain conditions is being evaluated, however, the efficacy is still not consistently established.

[Thesis Statement] In the present investigation, therefore, we discuss the different aspects related to cannabinoids and their implications in the management of chronic pain conditions. This review will also discuss the safety profile of the cannabinoids together with the legal considerations that hinder their use in different countries.

Fig. 4.3 A real example of an introduction with its sections labeled

4.4 Prevalent Pitfalls to Avoid

4.4.1 Insufficient Review of Existing Literature

An inadequate literature review can arise when the review is too concise, shallow, leaves out important studies, or fails to critically evaluate prior research. This shortcoming can weaken the basis of the study by not giving the essential context and rationale for the study.

Let us focus on practical applications:

Example with Insufficient Review of Existing Literature

> Previous studies have examined the link between exercise and mental health. This paper will explore this relationship in greater depth.

- **Problem**: This study lacks specificity, failing to adequately describe the existing research or explain how it relates to the present study.

Example with Sufficient Review of Existing Literature:
See link [4] under "Further Reading."

Powerful Tips
- **Be Thorough:** Review a wide variety of relevant studies to establish a comprehensive background to the research issue.
- **Be Specific:** Provide specific studies along with their methods, conclusions, and applicability to the ongoing research.
- **Be Critical**: Examine and assess the current research, highlighting its advantages, limitations, and areas needing further investigation.
- **Be Structured**: Organize the literature review in a coherent manner by classifying related studies according to themes or key findings.
- **Be Relevant**: Pay close attention to the studies that are most pertinent to the goals and research issue.

4.4.2 Inadequate Background Data

Definition When background data is inadequate, it means that the reader is not given enough context to comprehend the research issue and its impact.

Importance Background data establishes the context for the study by providing essential details about the subject, pertinent theories, earlier research, and important terminologies.

Inadequate Background Data May Result in
- **Audience Confusion:** Insufficient context makes it difficult for the audience to comprehend the impact of the study and how it relates to the larger scope of study.
- **Inadequate Rationale:** It can be challenging to defend the necessity and value of the study if there is insufficient prior data.
- **Misunderstanding**: Insufficient contextual information may cause the goals, procedures, and conclusions of the study to be misunderstood.

Let us focus on practical applications

Example of Inadequate Background Data

> Numerous academics have looked into how social media affects kids in recent years. This study investigates the connection between teen anxiety and social media use.

- **Problem**: The conceptual structure, important concepts, and prior research are not specifically covered in this introduction. The audience is not given adequate background data to comprehend the significance of the study.

Example of Adequate Background Data
See link [5] under "Further Reading."

Powerful Tips
- **Review-Related Literature:** Provide an overview of the most important research on your subject.
- **Provide Context**: Describe the larger background of your study issue.
- **Explain Keywords**: Make sure that any specialized words or ideas are explained in detail.
- **Determine the Research Gap**: Emphasize the aspects of your issue that are still unknown or unclear.
- **Be Brief**: Give the reader just enough background data without going into too much details.

4.4.3 Ignoring the Research Gap

Definition Ignoring the research gap entails not recognizing which particular part of the subject has not received enough attention or investigation in the current literature.

Importance Research gap is an essential element as it supports the novelty of the study. If this gap is not pointed up, the research might seem unnecessary.
Ignoring the research gap may result in:

- **Absence of Rationale**: If the study does not seem required or pertinent, its perceived worth will be lowered.
- **Redundancy**: The study can appear to be a repetition of previous research, providing no fresh perspectives or advancements in the subject. It is crucial to explain your reasoning, even if your approach is comparable to that of earlier research. For example, you may want to confirm your methodology in your sample population because it has not been utilized in many studies.
- **Lack of Interest**: If readers fail to recognize the research's distinctive contribution or goal, they may get disinterested.

Let us focus on practical applications:

Example of Ignoring the Research Gap:

> The impact of exercise on mental health has been the subject of numerous studies. This essay examines the connection between depression and physical activity.

- **Problem**: This introduction fails to emphasize the research's novel contribution by failing to identify the association between physical exercise and depression that has not been examined.

Example of Identifying the Research Gap
See link [6] under "Further Reading."

Powerful Tips
- **Perform a Comprehensive Literature Review**: Recognize what has been researched and where deficiencies exist by recognizing the momentous state of research in your field.
- **Be Specific**: Precisely state the gap of knowledge that has not been addressed in previous research.
- **Relate to Your study**: Describe how your study will fill this knowledge gap and advance the field.
- **Use Evidence**: Cite earlier research to back up your claim that there is a gap.
- **Highlight Significance**: Clarify how addressing this knowledge gap significantly advances academic understanding or practical applications.

4.4.4 Vague Research Goals and Purpose

The research introduction must include a clear description of objectives and scope since it establishes the goals of the study and the parameters for its execution.
Let us focus on practical applications:

Example of Vague Research Goals and Purpose

> The effects of climate change on agriculture are examined in this research.

- **Problem**: The general statement lacks specificity, failing to delineate the particular facets of climate change and agriculture to be investigated. Additionally, the geographical and temporal parameters are not clearly defined.

Example with Clear Goals and Purpose
See link [7] under "Further Reading."

Powerful Tips
- **Be Particular**: Give a clear explanation of the study's objectives and avoid using general or ambiguous language.
- **Identify Critical Aspects:** Focus on the principal domains or facets that will be addressed.
- **Demarcate Outlines:** Establish the geographical, temporal, and conceptual parameters of the research.
- **List Objectives**: Clearly state the precise questions or research goals that the study will try to answer.
- **Remain Realistic**: Make sure the goals and parameters are feasible within the limitations of the study.
- **Ensure Coherent Flow**: It is important to ensure the introduction flows logically. The thesis statement and objectives should be distinct yet complementary. The thesis presents the hypothesis or research question, while the objectives outline how this will be investigated. Avoiding repetition between these sections strengthens the introductory paragraph and sets the stage for the remainder of the paper.

4.4.5 Failure to Establish the Research Significance

This mistake could reduce the potential impact of the study by making it seem unimportant or inconsequential. Readers may find it difficult to appreciate the study and the reasons they should be interested in it if the research's significance is not clearly established.

Let us focus on practical applications:

Example of Failing to Convey Research Importance

> This study examines how teens' use of social media affects their sleep habits.

- **Problem**: There is no explanation of the need to research how social media affect sleep habits. The reader can question the significance of this study or its ramifications.

Example with Established Research Significance
See link [8] under "Further Reading."

Powerful Tips
- Connect the study topic to more general concerns or trends that emphasize its significance and applicability.
- Explain the Useful Implications: Talk about the possible advantages or practical uses of the study's conclusions.
- Fill in Knowledge Gaps: Determine which gaps in the body of literature need to be filled by the research.
- The research's possible influence on the field, society, or particular populations should be highlighted.
- To demonstrate the importance of research, give specific instances or situations.

4.4.6 Absence of a Clear Thesis Statement

Definition The thesis statement serves as the central claim or contention that directs the entire analysis. It provides a brief outline of the main idea or assertion of the work and is usually located at the conclusion of the introduction.

Importance
- The thesis statement not only provides clarity and focus on the research but also clearly conveys the study's purpose to the reader.
- Expert reviewers, who are knowledgeable in the subject, may evaluate the thesis statement alone without fully reading the introduction, making it crucial for the statement to be precise and well-crafted.

A lack of a strong thesis statement in a research introduction can lead to a number of problems:

- **Confusion**: The reader may struggle to understand the purpose and main argument of the paper. To avoid this, use precise terminologies from the recent literature and clearly define the key aspects of the relationship being investigated, including the study sample, outcome measures, and predictor variables.
- **Lack of Focus**: The paper risks becoming disorganized, straying into unrelated topics, and making the argument hard to follow. To maintain focus, limit the paper to 1–2 primary objectives. Any supplementary analyses should be mentioned in the discussion. As a best practice, aim to address one central question.
- **Weak Argumentation**: A well-defined thesis is essential for constructing strong, coherent arguments. Without it, the arguments may lack clarity and support, weakening the overall impact of the research.

Let us focus on practical applications.

In this article, I will address climate change.

- **Problem:** This statement lacks precision and is very general, making it difficult to discern a clear point or course of action.

Example of a Clear Thesis Statement
See link [9] under "Further Reading."

Powerful Tips
- **Be Specific:** Avoid using generic or ambiguous language and clearly convey the core point or argument.
- **For clarity and effect**, keep your argument to one or two phrases.
- **Provide Direction**: Highlight the key points or arguments that will guide the paper.
- **Revise as Needed**: Refine the thesis as your research develops and new insights emerge.

4.5 Overly Technical Language

Definition When jargon, complicated phrases, and highly specialized vocabulary are used excessively, it can be difficult for readers, especially those outside the field—to understand the information.

While technical language can be important in academic writing, using it too frequently in the introduction may lead to:

- **Reader Disengagement**: Excessive complexity can make the text feel intimidating or inaccessible, causing readers to lose interest.
- **Lack of Clarity:** Important concepts and the relevance of the research can become unclear due to excessively complex language.
- **Diminished Impact:** The introduction may fail to convey the importance of the research if readers struggle to understand its message.

Let us focus on practical applications:

Example of Overly Technical Language

> This research investigates the thought processes individuals use when learning a second language, particularly how different forms of memory knowledge-based and skill-based interact to facilitate understanding of sentence structure.

Problem This sentence is packed with specialized terminologies, such as "metacognitive strategies," "second language acquisition," "declarative and procedural memory systems," and "syntactic parsing." Such jargon can be overwhelming and confusing for readers who lack expertise in the field.

4.5.1 Neglecting the Audience

Definition Neglecting the audience means overlooking the readers' background, level of understanding, and interests when writing the introduction.

This can manifest itself in several ways, such as using excessive technical language for non-experts, providing insufficient background for newcomers to the topic, or not engaging the readers effectively.
Let us focus on practical applications:

Powerful Tips
- **Identify Your Audience** (e.g., experts or general readers) and adapt your language accordingly by referencing articles from target journals to match their expertise level.
- **Simplify Communication**: Use straightforward and accessible language, avoiding unnecessary technical terms unless they are essential and well-defined.
- **Offer Context:** Provide sufficient background information to help readers understand the importance and context of your research.
- **Capture Interest**: Start with an engaging introduction that highlights the relevance and significance of your research topic.
- **Address Potential Concerns:** Consider the questions or concerns your readers might have and address them effectively in the introduction.

Example with Simplified Language
See link [10] under "Further Reading."

Powerful Tips
- **Know Your Readers:** Tailor your language to suit your intended readers, ensuring it is accessible to both experts and general audiences.

- **Clarify Terms:** When using technical terminology is essential, it offers clear definitions or explanations.
- **Use Analogies:** Simplify complex concepts by incorporating relatable analogies or straightforward examples.
- **Limit Technical Language:** Avoid excessive use of specialized terms, particularly at the beginning.
- **Seek Feedback:** Invite input from colleagues or professionals outside your field to evaluate the clarity and accessibility of the introduction.

4.6 Poor Organization and Flow

Definition The introduction is hard to understand because of its poor organization and flow, which is defined as a lack of logical structure and coherence.

This may happen if concepts are presented randomly, if there are poor or nonexistent transitions between sections, and if the story is not coherent as a whole. A well-structured introduction should seamlessly lead the reader from the study's general background to its particular goals.

Let us focus on practical applications:

Example of Poor Organization and Flow

> Agriculture is impacted by climate change in a number of ways. The effect on crop production has been the subject of numerous research. The economic effects of these developments will be covered in this essay. Water availability will be impacted by the increased weather variability predicted by climate projections. Higher temperatures shorten the growing season for a variety of crops, according to research.

- **Problem**: The concepts are disorganized and lack obvious links. The discussion of the economic ramifications seems out of place, and the topics change abruptly.

Example with Good Organization and Flow
See link [11] under "Further Reading."

Powerful Tips
- **Make an Outline:** List the key topics you wish to discuss in the introduction before you start writing.
- **Consider an Inverted Triangle as an Example:** Introduce the fundamental ideas associated with your issue in a broad manner. You can introduce

increasingly focused subjects as you go forward with the introduction until you have sufficient data to support your thesis statement.
- **Incorporate Linking Phrases**: Utilize connecting phrases and sentences to ensure smooth transitions between ideas and sections.
- **Adhere to a Logical Order**: Information should be presented logically, starting with a basic backdrop and ending with specific goals.
- **Keep Your Attention:** Remain on topic and refrain from bringing up irrelevant subjects.
- **Edit for Coherence**: Go over and edit the introduction to make sure it makes sense and that every section adds to the story as a whole.

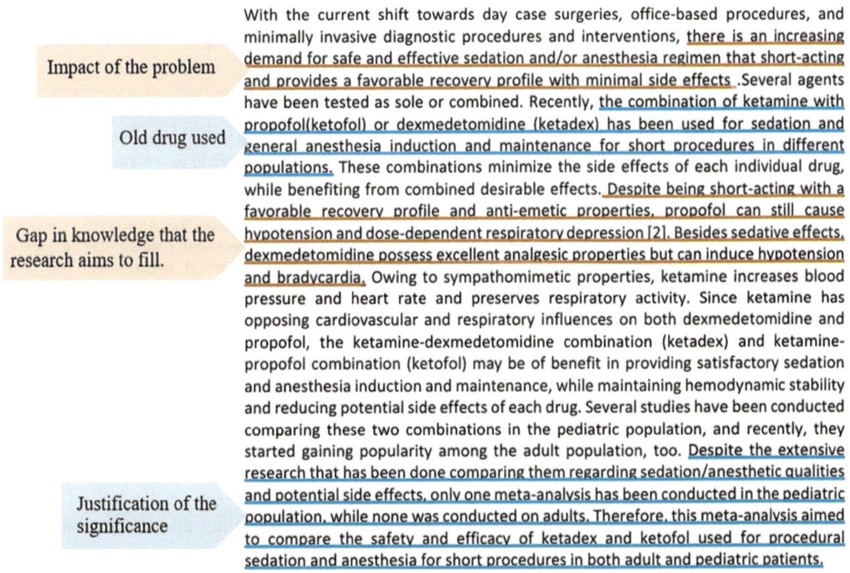

Impact of the problem: With the current shift towards day case surgeries, office-based procedures, and minimally invasive diagnostic procedures and interventions, there is an increasing demand for safe and effective sedation and/or anesthesia regimen that short-acting and provides a favorable recovery profile with minimal side effects. Several agents have been tested as sole or combined. Recently, the combination of ketamine with propofol(ketofol) or dexmedetomidine (ketadex) has been used for sedation and general anesthesia induction and maintenance for short procedures in different populations.

Old drug used: These combinations minimize the side effects of each individual drug, while benefiting from combined desirable effects. Despite being short-acting with a favorable recovery profile and anti-emetic properties, propofol can still cause hypotension and dose-dependent respiratory depression [2]. Besides sedative effects, dexmedetomidine possess excellent analgesic properties but can induce hypotension and bradycardia.

Gap in knowledge that the research aims to fill.: Owing to sympathomimetic properties, ketamine increases blood pressure and heart rate and preserves respiratory activity. Since ketamine has opposing cardiovascular and respiratory influences on both dexmedetomidine and propofol, the ketamine-dexmedetomidine combination (ketadex) and ketamine-propofol combination (ketofol) may be of benefit in providing satisfactory sedation and anesthesia induction and maintenance, while maintaining hemodynamic stability and reducing potential side effects of each drug. Several studies have been conducted comparing these two combinations in the pediatric population, and recently, they started gaining popularity among the adult population, too. Despite the extensive research that has been done comparing them regarding sedation/anesthetic qualities and potential side effects, only one meta-analysis has been conducted in the pediatric population, while none was conducted on adults.

Justification of the significance: Therefore, this meta-analysis aimed to compare the safety and efficacy of ketadex and ketofol used for procedural sedation and anesthesia for short procedures in both adult and pediatric patients.

Fig. 4.4 A real example illustrating key points to include when writing an introduction for a drug

	Low back pain (LBP), defined as a disorder of the lumbosacral spine and categorized as acute, subacute, or chronic, can be a debilitating condition for many patients. <u>While
Statistical facts capture interest and emphasize significance	the majority of patients with LBP are acute and self-limited, up to 40% of patients can develop pain lasting more than 6 weeks.</u> Attributable to various etiologies, LBP can often be persistent and difficult to treat, potentially leading to increased healthcare utilization, disability, and lost wages. <u>While there is a multitude of pharmacologic, psychological, and physical/rehabilitative treatment options, the use of
The old treatment	complementary alternative medicine (CAM) is becoming popular around the world.</u> Global studies have shown that up to 40–55% of patients are seeking out CAM to treat their LBP, with treatments including chiropractic manipulation, massage therapy, acupuncture, sleep support, and other non-traditional therapies. Acupuncture is traditionally utilized as a therapy in eastern medicine, where acupuncturists insert needles to mechanically or electrically stimulate specific anatomical points along classically defined meridians to relieve illnesses and maladies. <u>In the USA, based on a
Specified population	2012 survey, an estimated 3.8 million adults utilized acupuncture within the last year to treat pain and other ailments.</u> With acupuncture growing in popularity within western medicine, physicians may find it difficult to make adequate recommendations regarding this alternative therapy. <u>This evidence-based systematic
Thesis statement	review will focus on the use of acupuncture and its role in the treatment of LBP to help better guide physicians in their practice.</u>

Fig. 4.5 A real example illustrates key points to include when writing an introduction for an intervention

	Currently, left atrial appendage occlusion (LAAO) or left atrial appendage closure has emerged as an alternative treatment to prevent stroke in patients with chronic non valvular atrial fibrillation(NVAF) (Reddy et al., 2011) who are unable to undergo oral anti-coagulation (OAC) therapy due to a formal contraindication for OAC or an unacceptable risk of bleeding (Kirchhof et al., 2016).The procedure is mostly performed with fluoroscopic or trans-esophageal echocardiography (TEE) guidance (Wunderlich et al.,2015). <u>However, TEE requires general anesthesia or profound
Significance	sedation. Using intracardiac echocardiography (ICE) to guide LAAO, which only requires conscious sedation, may have potential advantages over TEE.</u> Although some studies (Aguirre et al., 2018; Frangieh et al., 2017; Hemam et al., 2019; Ho et al.,2007; Masson et al., 2015; Matsuo et al., 2016) with small samples of patients demonstrated that the feasibility and efficacy of Ice-guided
Gap in the existing literature	LAAO were not inferior to those of TEE, <u>the findings remain inconsistent, and few meta-analyses have assessed the results. Therefore, we sought to systematically
Aim of the study	evaluate the extent to which ICE is an alternative to TEE.</u>

Fig. 4.6 A real example illustrating key points to include when writing an introduction for an imaging technique

4.7 Real Examples

The introduction differs according to the topic of the research paper.

Let us see how things go by real examples of common topics (Drugs, Intervention, and Imaging Techniques).

- Drugs [5].
  As shown in Fig. 4.4.
- Intervention [10].
  As shown in Fig. 4.5.
- Imaging Techniques [11].

As shown in Fig. 4.6.

4.8 Conclusion

You may write an opening that not only captures readers' interest but also lays the groundwork for an engaging and influential research paper by adhering to these rules and steering clear of typical errors.

4.9 Final Tips

- To make sure your introduction is coherent and clear, go over it several times.
- To find areas that need work, ask advisors, mentors, or peers for their opinions.
- Consider your audience and adjust your language and content to suit their requirements and preferences.
- Remain committed to your research goals and make sure that each section of your introduction advances them.
- Have faith in the importance of your study and its possible influence on your community or field.

4.10 Note: AI Tools

- **Jenni AI**
  Your path to academic success can be accelerated and your academic writing experience enhanced with Jenni AI.
  https://jenni.ai/for-researchers
- **paperpal**
  With real-time, topic-specific language recommendations that help you write better, faster, you can increase your chances of success. https://paperpal.com/
- **avidnote** https://avidnote.com/
- **perplexity.ai**
  Your AI-powered research & collaboration hub customized for your purpose
  https://www.perplxity.ai/spaces

Further Reading
Cannabinoids and Their Role in Chronic Pain Treatment: Current Concepts and a Comprehensive Review [4]

The Possible Application of Ketamine in the Treatment of Depression in Alzheimeros Disease [5]

Anesthetic Considerations for Cesarean Delivery After Uterine Transplant [6]

Efficacy and Safety of Ketamine-Dexmedetomidine Versus Ketamine-Propofol Combination for Periprocedural Sedation: A Systematic Review and Meta-analysis. [7]

Ketamine: Pro or antiepileptic agent? A systematic review [8]
Intraoperative Hypotension Increased Risk in the Oncological Patient [9]
Mitral Regurgitation in Patients Undergoing Noncardiac Surgery [10]
Con: Routine Use of the Hypotension Prediction Index in Cardiac, Thoracic, and Vascular Surgery [11]

References

1. Caulfield J. Writing a research paper introduction | step-by-step guide. Scribbr Published September 5, 2024. https://www.scribbr.com/research-paper/research-paper-introduction/
2. Alex D, Alex D. How to write a research paper introduction (with examples). Paperpal Blog Published September 18, 2024. https://paperpal.com/blog/researcher-resources/how-to-write-a-research-paper-introduction-with-examples
3. Hassan M. Research paper introduction – writing guide and examples. Research Method Published November 12, 2024. https://researchmethod.net/research-paper-introduction/
4. (PDF) Cannabinoids and their role in Chronic pain treatment: Current concepts and a Comprehensive review. ResearchGate. https://www.researchgate.net/publication/364254149_Cannabinoids_and_Their_Role_in_Chronic_Pain_Treatment_Current_Concepts_and_a_Comprehensive_Review
5. (PDF) Efficacy and Safety of Ketamine-Dexmedetomidine versus Ketamine-Propofol combination for periprocedural sedation: a systematic review and meta-analysis. ResearchGate. https://www.researchgate.net/publication/377358162_Efficacy_and_Safety_of_Ketamine-Dexmedetomidine_Versus_Ketamine-Propofol_Combination_for_Periprocedural_Sedation_A_Systematic_Review_and_Meta-analysis
6. (PDF) Ketamine: Pro or antiepileptic agent? A systematic review. ResearchGate. https://www.researchgate.net/publication/377300540_Ketamine_Pro_or_antiepileptic_agent_A_systematic_review
7. (PDF) Intraoperative hypotension increased risk in the oncological patient. ResearchGate. https://www.researchgate.net/publication/350018351_Intraoperative_Hypotension_Increased_Risk_in_the_Oncological_Patient
8. Mitral regurgitation in patients undergoing noncardiac surgery | Request PDF. ResearchGate. https://www.researchgate.net/publication/354292951_Mitral_Regurgitation_in_Patients_Undergoing_Noncardiac_Surgery
9. (PDF) Con: Routine use of the Hypotension Prediction Index in cardiac, thoracic, and vascular surgery. ResearchGate. Published September 1, 2023. https://www.researchgate.net/publication/345657778_Con_Routine_Use_of_the_Hypotension_Prediction_Index_in_Cardiac_Thoracic_and_Vascular_Surgery
10. (PDF) Acupuncture for the management of low back pain. ResearchGate. Published August 1, 2020. https://www.researchgate.net/publication/348486142_Acupuncture_for_the_Management_of_Low_Back_Pain
11. Left Atrial Appendage Occlusion: Transesophageal echocardiography versus Intracardiac echocardiography—PRO: Intracardiac echocardiography | request PDF. ResearchGate Published March 1, 2020. https://www.researchgate.net/publication/370563589_Left_Atrial_Appendage_Occlusion_Transesophageal_Echocardiography_versus_Intracardiac_Echocardiography_-_Pro_Intracardiac_Echocardiography

Discussion

5

Alaa Abdeltawab Abouammar, Rodaina Ehab Ashour, Islam Mohammad Shehata, Omar Viswanath, and Natalie Strand

> *If you cannot explain it simply, you do not understand it enough*
> – Albert Einstein

5.1 Definition

The discussion section is the core of any research paper. It must be structured in such a way that it provides a concise and well-summarized interpretation as well as an explanation of the results.

A successful discussion section will highlight the results for the reviewer and reader and place emphasis on their significance in the current existing body of literature [1].

5.2 Role of Discussion (Fig. 5.1)

A. A. Abouammar (✉) · R. E. Ashour
Faculty of Medicine, Modern University for Technology and Information, Cairo, Egypt
e-mail: alaa.94685@medicine.mti.edu.eg; Rodaina.95749@medicine.mti.edu.eg

I. M. Shehata
Department of Anesthesiology, Faculty of Medicine, Ain Shams University Cairo, Cairo, Egypt
e-mail: Islam.shehata@med.asu.edu.eg

O. Viswanath
Department of Anesthesiology, Creighton University School of Medicine, Phoenix, AZ, USA

Mountain View Headache and Spine Institute, Phoenix, AZ, USA

N. Strand
Department of Anesthesiology, Division of Pain Medicine, Mayo Clinic Arizona, Phoenix, AZ, USA
e-mail: Strand.Natalie@mayo.edu

© The Author(s), under exclusive license to Springer Nature Switzerland AG 2025
I. M. Shehata, O. Viswanath (eds.), *How to Successfully Publish a Manuscript*,
https://doi.org/10.1007/978-3-031-92538-2_5

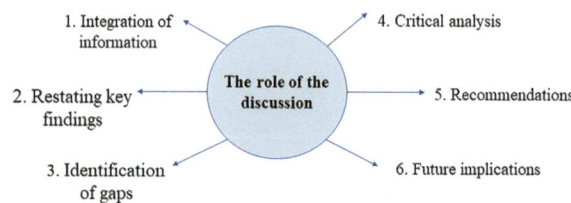

Fig. 5.1 Key roles of the discussion section in scientific writing [2]

5.3 How to Prepare for the Discussion

1. Before you write, you should:
   - Read more than ten papers.
   - Read actively (take notes), not passively.
   - Draw diagrams and mind maps as you read along.
   - Ask yourself questions as you go along. "Does my study agree with this? How might these ideas or concepts be used in my writing?"
2. Planning your writing:

   Planning should be systematic. It will be reflected in your writing as it ensures clarity about the questions/issues to be addressed, and this is done by the following:
   1. Focus your thoughts and outline the data to ensure that you are telling a coherent story.
   2. Draft an action plan, list all the tasks, and define certain milestones.
   3. Draw diagrams, mind maps, or abstracts.
   4. Create a potential timeline.
   5. Thoroughly revise your draft. Do not fall into the error of thinking that you can get it right the first time and without the help and several stages of drafting and redrafting [3].

5.4 General Writing Rules

The main key to writing a comprehensive discussion is to read because reading is the teacher of writing. We advise you to read at least 10 papers about your topic before writing your discussion section. Reading should help you to meticulously understand the different aspects of your topic and discover various points of view. This should help you find the knowledge gap and ultimately form your own opinion [4]. A summary of key recommendations is provided in Fig. 5.2, which outlines essential Do's and Don'ts for effective academic writing.

Do's	Don'ts
Make sure to checkout each journal guidelines before submitting your article.	Avoid bias by ignoring results and only considering those supporting your research question.
Write shortly and go for maximum clarity and conciseness.	Don't rewrite the results but interpret and analyze.
It's essential to acknowledge the work of others properly.	Don't overstate the impact of your findings. Don't use (Must), but use (could, can, may).
Always get other people to read your work frequently before submitting.	Avoid using unsupportive evidence or counterevidence.
Use active voice rather than passive voice for a more engaging reading.	Don't shift back and forth between tenses. Use the present tense to write your results, and past tense to show past results.

Fig. 5.2 Table summarizing the general writing rules for scientific articles

5.5 Structure of the Discussion Section

5.5.1 How to Break Down Your Body into Sections?

It is important to divide your body into subtitles, starting with the basic to advanced level (you should satisfy all levels of readers). Subheadings for the sections of your review will assist you in creating a logical structure and keep you focused on subtopics relevant to your research aim [5]. As illustrated in Fig. 5.3, subtitles should be concise, relevant, and aligned with the content of the section.

Questions that could facilitate writing a subtitle:

1. What key idea needs further focus, discussion, or elaboration?
2. What practical solutions can address this issue?
3. What fresh or innovative perspective does this paper introduce?
4. How does this study or work contribute to addressing the problem?

Examples of Subtitles
Your paper could be about one of these four general topics: A drug, a technique (surgical or nonsurgical), an imaging modality, or a disease. To better understand how to create effective subtitles, Fig. 5.4 provides examples of subtitles tailored to different topics [6].

N.B. It is important to note that these are just examples and you are free to tailor them according to your needs.

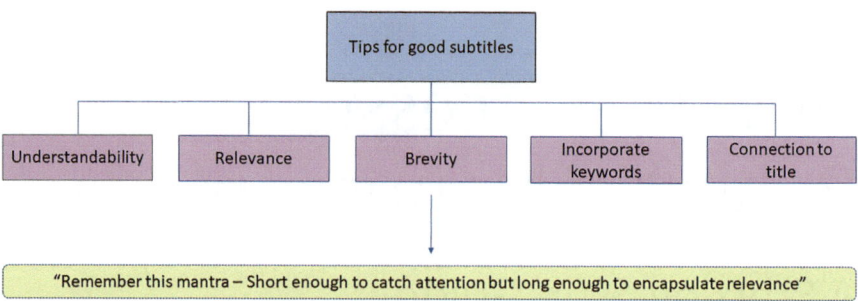

Fig. 5.3 Diagram outlining tips for crafting good subtitles

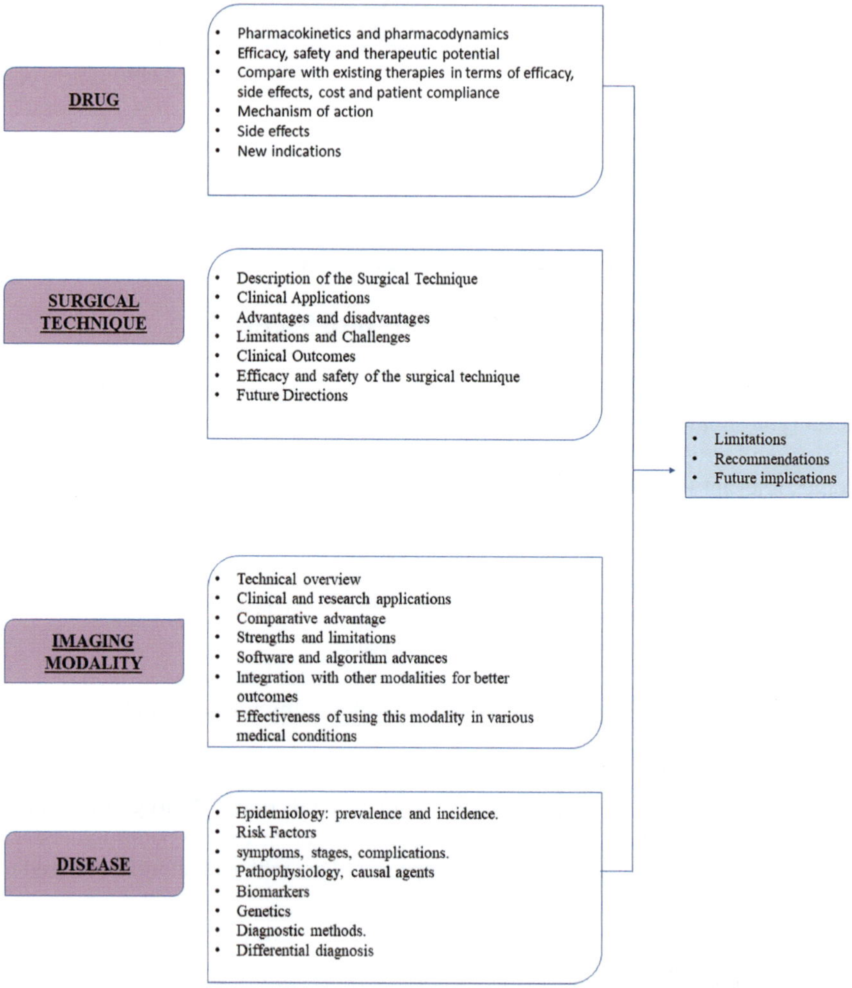

Fig. 5.4 Examples of subtitles tailored to different topics, demonstrating how to align subtitles with the focus of the scientific content

5.6 Structure of the Subtitle

5.6.1 Constructing a Paragraph

The paragraph is the building unit of the subtitle. The paragraphs of the discussion emphasize the comparison of the findings with other works. This comparison demonstrates the support and opposition to your results. Supporting your findings is crucial to highlighting the significance of your study's results. In these paragraphs, your task is to interpret and analyze the most important results effectively. The study's findings are prioritized based on their importance, *and a paragraph is constructed for each finding. So, we provide a step-by-step guide, as shown in Fig. 5.5, on how to construct a good one* [7].

Each paragraph should include:

5.6.2 An Introductory Sentence

Focus on the core of the paragraph, grasping the reader's attention. Different ways to write your introduction sentence are as follows:

- Highlight the importance of the topic.
- Briefly start with a broader view of the topic.
- Pose a compelling question or problem.
- An answer to "What have we found?"

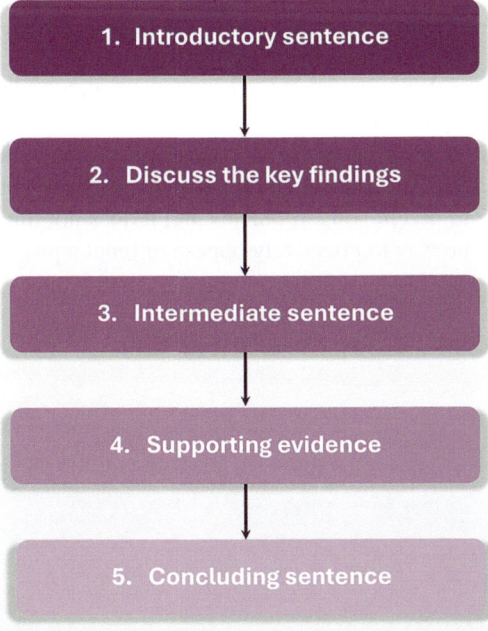

Fig. 5.5 Flowchart illustrating a general scheme for constructing a paragraph, ensuring a logical flow

1. **An Intermediate Sentence.**
   - The main idea of the subtitle should be emphasized: mention and compare existing results.
   - Restate the key insights of the study using "constrained" and "not too strongly assertive" statements.
   - Mention the results of other papers: the previous studies can provide context or support your findings.
2. **A Concluding Sentence.**
   Apply topic closers:
   - Use conclusionary verbs such as "suggests, demonstrates, shows, makes clear that, illustrates, informs us that, implies, and confirms."
   - Begin with a transitional word (therefore, thus, as such, as a result, consequently, accordingly, and evidently), informing the reader that a conclusion is about to be drawn. Highlight previous information, such as "This approach" means that "the previous information" is now going to receive some final remark [8].

5.6.3 How to Construct an Argument?

Although organizing your evidential material is your choice, there are several common patterns for you to consider, as shown in Fig. 5.6. McCarthyPM et al. 2018 [9]. presented three of these formationsin Fig. 5.6.

Q&A
Q. Which comes first, the evidence or the counterevidence?
What is most likely to convince your audience will determine this. Start with the thesis statement if your topic is one that not many people are familiar with. Start with the counterevidence if your audience is already aware of the issue and argument.
Q. What if there are no refutations or rebuttals?
A. If all you have to do is demonstrate that your stance is important, you just have to accept some problems and it does not imply that you are incorrect if there is no way to effectively oppose or rebut a piece of counterevidence [9].

5.6.4 Citing Results (Fig. 5.7)

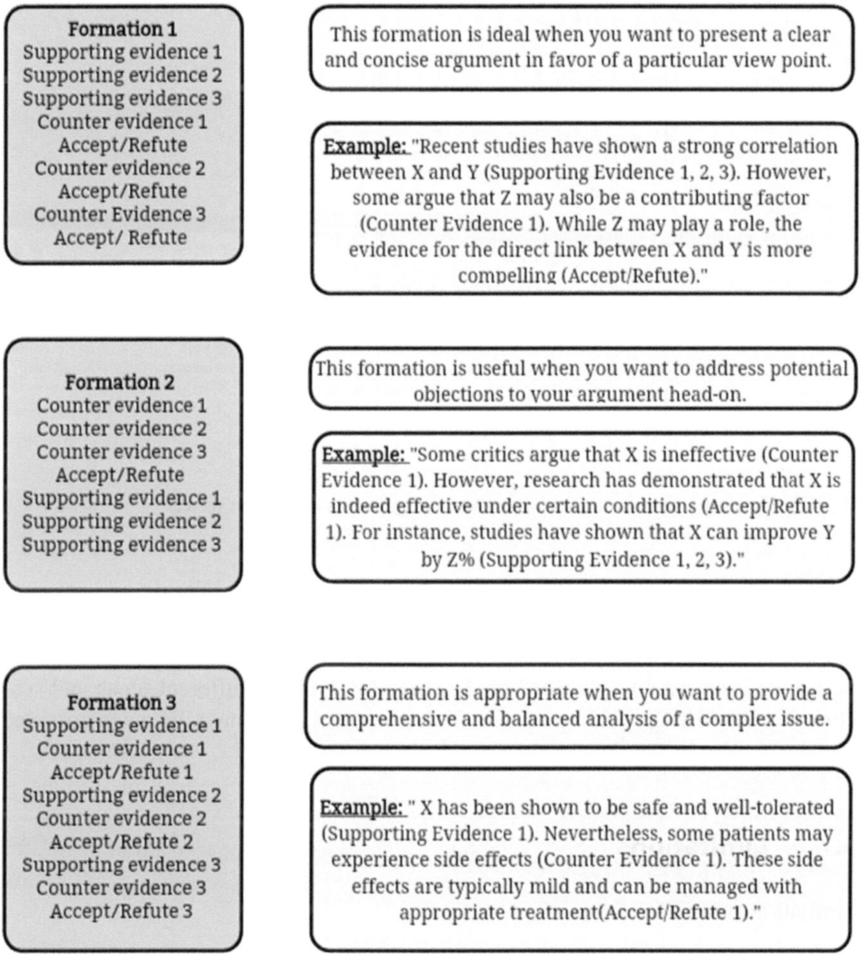

5 Discussion

Formation 1
Supporting evidence 1
Supporting evidence 2
Supporting evidence 3
Counter evidence 1
Accept/Refute
Counter evidence 2
Accept/Refute
Counter Evidence 3
Accept/ Refute

This formation is ideal when you want to present a clear and concise argument in favor of a particular view point.

Example: "Recent studies have shown a strong correlation between X and Y (Supporting Evidence 1, 2, 3). However, some argue that Z may also be a contributing factor (Counter Evidence 1). While Z may play a role, the evidence for the direct link between X and Y is more compelling (Accept/Refute)."

Formation 2
Counter evidence 1
Counter evidence 2
Counter evidence 3
Accept/Refute
Supporting evidence 1
Supporting evidence 2
Supporting evidence 3

This formation is useful when you want to address potential objections to your argument head-on.

Example: "Some critics argue that X is ineffective (Counter Evidence 1). However, research has demonstrated that X is indeed effective under certain conditions (Accept/Refute 1). For instance, studies have shown that X can improve Y by Z% (Supporting Evidence 1, 2, 3)."

Formation 3
Supporting evidence 1
Counter evidence 1
Accept/Refute 1
Supporting evidence 2
Counter evidence 2
Accept/Refute 2
Supporting evidence 3
Counter evidence 3
Accept/Refute 3

This formation is appropriate when you want to provide a comprehensive and balanced analysis of a complex issue.

Example: " X has been shown to be safe and well-tolerated (Supporting Evidence 1). Nevertheless, some patients may experience side effects (Counter Evidence 1). These side effects are typically mild and can be managed with appropriate treatment(Accept/Refute 1)."

Fig. 5.6 Examples of three possible formations to organize your evidence and counterevidence

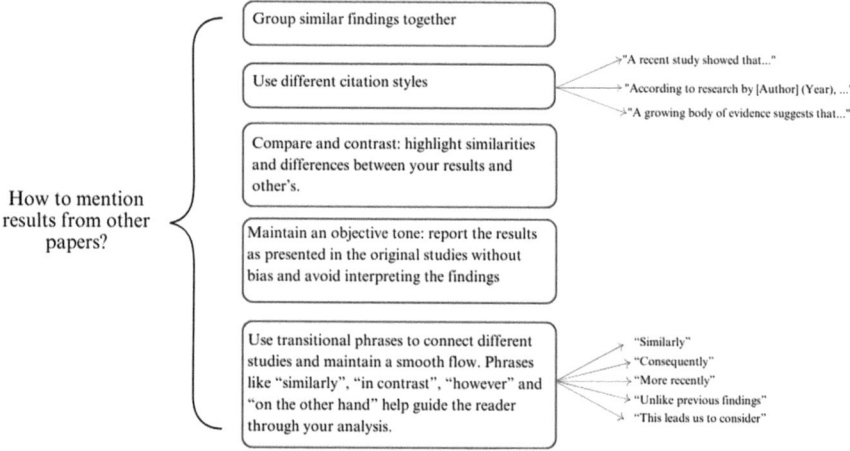

Fig. 5.7 General notes to consider when mentioning results from other papers

5.6.5 Organizing the Discussion Using Transition Words and Phrases

Transitional words and phrases help to smoothly connect different ideas and sections of your writing, guiding readers from one point to the next. We have provided some examples in Fig. 5.8.

5.7 Limitations

Limitations may arise from:

- A limited focus on a specific aspect of a broader topic.
- Authors may not be able to access all relevant databases or may miss key studies leading to superficial analysis.
- Language and accessibility barriers can limit the scope of the review.
- Methodological limitations as variability in study designs (e.g., randomized controlled trials vs. observational studies) can complicate comparisons and the synthesis of findings. Also, Quality of Studies as low-quality studies can undermine the overall reliability of the review's conclusions.
- Suggestions for further research can lead directly to future recommendations [10].

5 Discussion

Fig. 5.8 List of transition words and phrases

5.8 Recommendations

Based on the discussion of your results, you can offer suggestions for practical applications or future research. Avoid simply stating that more studies are necessary; instead, provide specific suggestions on how future work can expand on the aspects your research could not cover.

Examples: Recommendation sentence starters:

1. Additional research is required to determine…
2. Future investigations should consider…
3. Potential directions for future research include… [11]

5.9 Future Implications

Describe how your findings might affect the research topic. For example, your findings might point to gaps in the existing body of literature that future research could address. Studies to build on your findings might be required, or more study might be required to confirm your conclusions [12].

5.10 AI Tools

The following is a list of AI tools that can help you write a proper discussion section by ensuring accurate grammar and clarity in your writing.

- **Jenni AI**
  Jenni AI can elevate your academic writing experience and accelerate your journey toward academic excellence.
  Website: https://jenni.ai/for-researchers
- **Paperpal**
  Boost your chances of success with real-time, subject-specific language suggestions that help you write better, faster!
  Website: https://paperpal.com/
- **Avidnote**
  Website: https://avidnote.com/
- **Perplexity.ai**
  Your AI-powered research & collaboration hub customized for your purpose.
  Website: https://www.perplexity.ai/spaces

References

1. Taherdoost H. How to write an effective discussion in a research paper; a guide to writing the discussion section of a research article. Open Access J Addict Psychol. 2022;5:5. https://doi.org/10.33552/OAJAP.2022.05.000609.
2. San José State University Writing Center, Dunton R, Phillips E, Lenarz E, Ruprecht E, Weiskopf E, McCann E, Renecker H, Hoy E. Discussion section for research papers [Internet]. Fall. 2021:1–5.
3. Lantsoght EOL. Honing your academic writing skills. Springer; 2018.
4. Shon PC. The quick fix guide to academic writing: how to avoid big mistakes and small errors. Sage Study Skills; 2018.
5. Winter T. The power of Headings & Subheadings: tips to improve your writing [internet]. SEOwind. 2024.
6. Shehata I, Kohaf N, Elsayed M, Latifi K, Aboutaleb A, Kaye A. Ketamine: pro or antiepileptic agent? A systematic review. Heliyon. 2024;10 https://doi.org/10.1016/j.heliyon.2024.e24433.
7. Kumar A. Writing the discussion: the analysis should speak for itself. J Indian Soc Periodontol. 2022;26(5):421–2. https://doi.org/10.4103/jisp.jisp_317_22.
8. Şanlı Ö, Erdem S, Tefik T. How to write a discussion section? Turk J Urol. 2013;39(Suppl 1):20. https://doi.org/10.5152/tud.2013.049.
9. McCarthy PM, Ahmed K. Writing the research paper. London: Bloomsbury Academic; 2018.
10. Hassan M. Limitations in research - types, examples and writing guide [internet]. Research Method. 2024; [cited 2024 Nov 29]. Available from: https://researchmethod.net/limitations-in-research/
11. McCombes S. How to write a discussion section I tips & examples [internet]. Scr Theol. 2023; [cited 2024 Nov 29]. Available from: https://www.scribbr.com/dissertation/discussion/
12. Refnwrite. Discussion section examples and writing tips [internet]. Ref-n-Write: Scientific Research Paper Writing Software; 2023.

Conclusion and Abstract

6

Tabia Imtiyaz Khan, Islam Mohammad Shehata ⓘ,
Omar Viswanath ⓘ, and Giustino Varrassi

6.1 Conclusion

In research, a conclusion is not the end; it's the beginning of a deeper inquiry. – John Dewey

6.1.1 Definition

- Simply put, the conclusion is equivalent to the "**take home message**" of your paper [1].
- It goes beyond a mere summary; it combines the key points and significance: the implications of your findings.
- It demonstrates to the reader how you addressed a knowledge gap and outlines potential avenues for further exploration of the topic.

6.1.2 Importance [2, 3] (Fig. 6.1)

T. I. Khan (✉)
Faculty of Medicine, Ain Shams University, Cairo, Egypt

I. M. Shehata
Department of Anesthesiology, Faculty of Medicine, Ain Shams University Cairo, Cairo, Egypt
e-mail: Islam.shehata@med.asu.edu.eg

O. Viswanath
Department of, Anesthesiology, Creighton University School of Medicine, Phoenix, AZ, USA

Mountain View Headache and Spine Institute, Phoenix, AZ, USA

G. Varrassi
Department of Research, Fondazione Paolo Procacci, Roma, Italy

© The Author(s), under exclusive license to Springer Nature Switzerland AG 2025
I. M. Shehata, O. Viswanath (eds.), *How to Successfully Publish a Manuscript*,
https://doi.org/10.1007/978-3-031-92538-2_6

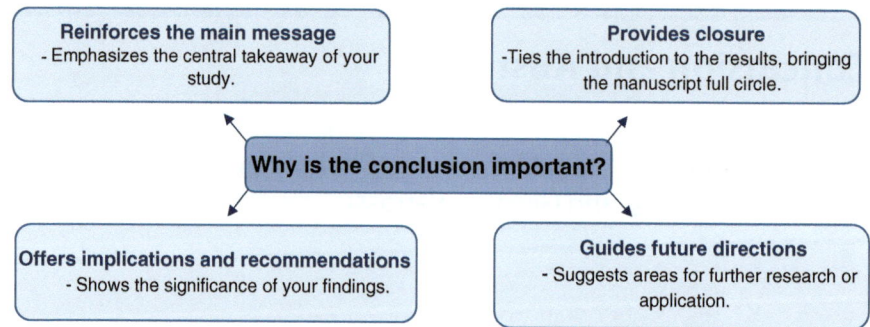

Fig. 6.1 Importance of conclusion

6.1.3 How to Write the Conclusion?

Writing a conclusion can be as difficult as writing your introduction.

In the Introduction, you start by identifying the research question or knowledge gap and then narrow down to your paper's focus. However, in the Conclusion, you summarize key findings and broaden the discussion to their larger implications.

For most papers, one well-written and developed paragraph is sufficient for a conclusion, although in some cases, a two- or three-paragraph conclusion may be required [2].

In general, the length of your conclusion should reflect the depth and scope of your paper's content.

6.1.3.1 Restating the Introduction
– Although it may sound repetitive, this **reinforces the main objective** of your paper for the reader [1].
– Rather than repeating the introduction verbatim, rephrase the research problem in a fresh and insightful manner.

6.1.3.2 Summarize Your Findings or Arguments
– **Restate the Key Findings:** Briefly summarize the main results or discoveries from your research, directly addressing the research questions or hypotheses.
– Focus on the **most important outcomes**, the insights, evaluations, and implications rather than detailed data.
– Tie the summary back to the purpose of your research. Show with evidence from the results section how your findings answer the initial research question or problem you set out to explore and that you have actually bridged a gap in knowledge [4].
– If your paper started with hypotheses, explain whether they were supported or not and discuss their significance. Do not worry if your results are negative or positive; they are still significant.

6.1.3.3 Discuss the Significance and Implications
– Reflect on the **"so what?"** aspect of your paper.
– Discuss the **practical implications**—in what ways could your findings influence future research, policy, or real-world applications?

6 Conclusion and Abstract

Table 6.1 Dos and don'ts while writing the conclusion

Dos	Don'ts
Keep it concise	Do not dwell on the methodology that should be in the body of the papers
Focus briefly on the results and main findings. Emphasize the significance and impact	Do not simply repeat your results or the discussion
Keep the language simple and clear	Do not use redundant phrases like "in conclusion" or "in summary"
Mention the limitations of your study and any negative results [2]	Do not introduce any new information

- Discuss the **broader impact**—provide recommendations for future research in your field. What questions or topics remain unresolved or deserve further exploration? [3]
- This demonstrates the significance of your findings and that your work paves the way for continued investigation.

6.1.4 Dos and Don'ts While Writing the Conclusion

6.1.5 Reviewing an Example for a Conclusion [5]

"Although there is currently no cure for AD, there are several approved medications and targets for drug therapy in clinical trials for symptomatic treatment. The FDA-approved medications for AD are acetylcholinesterase inhibitors and NMDA antagonists. Recent clinical findings in the last twenty years suggest that the nonselective NMDA antagonist ketamine may be beneficial in providing both neuroprotection and reduction of the neuropsychiatric symptoms in AD. Ketamine may prove to be more beneficial to patients than the standard treatments for AD because it has fewer side effects than acetylcholinesterase inhibitors and more of a broad mechanism of action than the NMDA antagonist, memantine. As a well-known analgesic and anti-inflammatory drug, it acts quickly, has long-lasting effects, and improves psychiatric symptoms with a smaller therapeutic dose than other medications. Ketamine has already proven successful in the treatment of psychiatric symptoms, specifically for treatment-resistant depression. This is important because depression may occur prior to memory loss in AD. In addition, ketamine is less likely to worsen cognition compared to other treatments like ECT for severe depression. Clinical trials have demonstrated that periodic doses of ketamine reduce symptoms such as suicidal ideation and psychosis. However, ketamine can also cause side effects such as dissociation, memory loss, and confusion. Based on these known side effects, the effect on individuals with dementia and depression may simply be explained as a drug side effect rather than a definitive improvement in symptoms. Additionally, vitals should be monitored closely when treating individuals with ketamine because individuals may experience an increase in blood pressure. This is a factor to consider before treating individuals with ketamine, especially in older individuals with cardiac comorbidities. Ketamine has been shown to act on the cellular level in opposing neurotoxins and protecting glial and neuronal function from the deleterious effects of inflammatory cytokines. Based on present information, we can summarize that ketamine may have a role in neuroprotection and the improvement of psychiatric symptoms in AD. Further research and clinical trials are warranted to prove or disprove this theory, but it is well worth investigating for a potential chance at improving the quality of life for millions of individuals."	- **Restates the introduction** (knowledge gap) to give context. -Summarizes the **key findings**: Recaps the key findings on ketamine's potential in AD treatment. -Discusses the **practical implications** through its faster action, fewer side effects, and psychiatric benefits, including its efficacy for depression, a common precursor to memory loss in AD. -Offers a **balanced perspective**: Acknowledges benefits (neuroprotection, psychiatric improvements) while addressing risks (side effects, blood pressure concerns), enhancing credibility. -Offers a **recommendation** for patients based on the results -Provides recommendation for future research and clinical trials.

6.2 Abstract

The abstract serves as the bridge between your research and its audience. It must make the complex simple, and the simple intriguing. – Professor Daniel J. Levitin

6.2.1 Definition

- The abstract is an originally written summary, not an excerpt, that describes the focus of your entire research manuscript [6].
- It condenses all the necessary information from the whole paper into a word-limited paragraph.

6.2.2 Importance [6, 7] (Fig. 6.2)

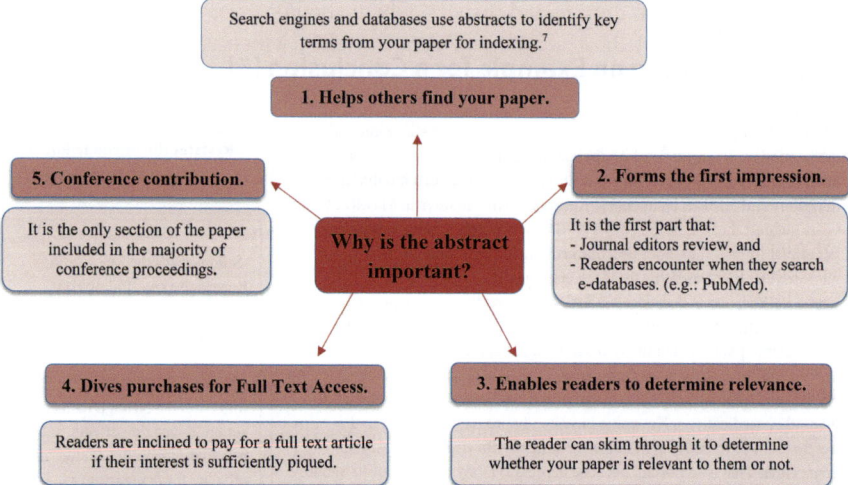

Fig. 6.2 Importance of abstract

6.2.3 When Should You Write the Abstract?

- Although the abstract appears as the first part of your paper, it is better to write it after you have completed the full paper [8].
- This allows you to distill the most important points and ensure the abstract is consistent with them.

6.2.4 Where to Place It in an Academic Paper?

- The abstract should be on its own page, and generally goes after the title page and acknowledgments, but before the table of contents.

6.2.5 Types of Abstracts [7–10]

Before drafting the abstract, you should determine which type of abstract is most appropriate for your manuscript and ensure that you are familiar with conventional guidelines for that abstract type.

Table 6.2 Informative vs descriptive abstract

Informative abstract	Descriptive abstract
Serves as a **concise summary** of the paper, **encompassing all key elements**	Serves as a **general overview** of the study without detailing the results or specifics
It is usually structured and contains the problem statement, methods, results, conclusion, and recommendations	It outlines the objectives, scope, and purpose while excluding specific results or recommendations
Suited for detailed, empirical studies, and reviews with clear findings and conclusions	Best suited for papers with interesting or complex research questions with more abstract findings
Examples: Journal and conference submissions, particularly in science, engineering.	Examples: Humanities and social sciences, and even theoretical papers in science

Table 6.3 Structured vs unstructured abstract

Structured abstract	Unstructured abstract
Divided into distinct specific sections (introduction, methods, results, and discussion) (IMRAD structure)	A single continuous paragraph without predefined sections
Provides a clear and detailed overview, making it easier to find specific information	Provides a general summary, which offers more flexibility in presentation
Commonly used and recommended (ICMJE 1993) for scientific and medical and clinical papers	Often used in humanities, social sciences, some conference abstracts, and scientific papers like case reports and narrative reviews

6.2.6 How to Choose the Right Type of Abstract for Your Submission?

1. **Follow submission guidelines:** Always comply with the journal or conference submission instructions, which will often specify the required abstract type and any content or organizational rules that may apply.
2. **Read published articles:** Review abstracts from articles already published in your target journal to understand their preferred style and format.
3. **Familiarize yourself with the word count:** Most journals specify a word count for abstracts, typically between 250 and 350 words, although this varies by journal.
   – Keep in mind that some databases may have word limits for abstracts [8].

6.2.7 Structure of an Abstract (Fig. 6.3) [11]

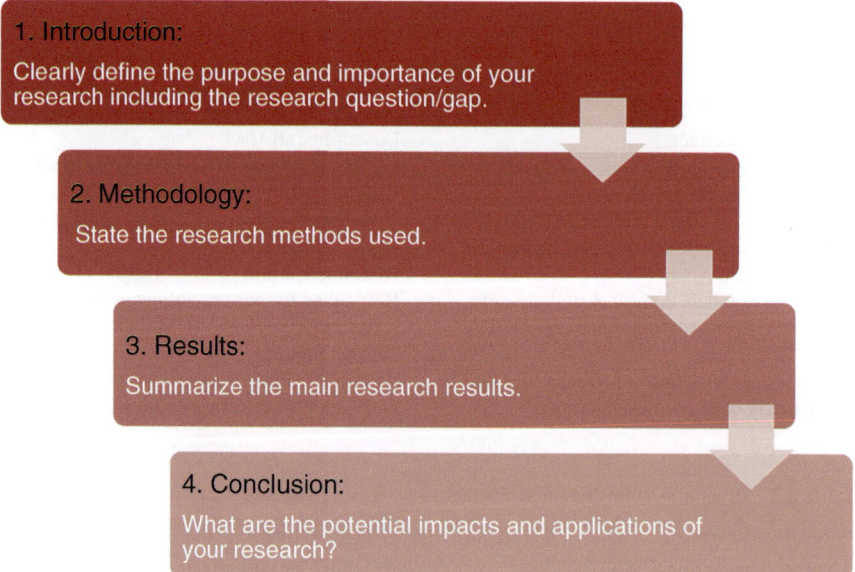

Fig. 6.3 Structure of abstract

6 Conclusion and Abstract

6.2.8 How to Write the Abstract?

A framework has been developed for this chapter where the most relevant types of journal articles/papers [12] have been **categorized into four types**.

- Each type has distinct expectations for what the key focus of the abstract should be, along with the introduction, methodology, results, and conclusions of it.
- This framework ensures that there is a focused, relevant content for each section.

6.2.8.1 Introduction

Table 6.4 Categorization of journal articles with key abstract focus areas

	Original research	Review articles	Clinical application and practice	Commentary, opinions, and methodological advances
Overview	Focuses on studies that create new data or findings	Summarizes and interprets existing research to draw conclusions or provide recommendations.	Targets articles that influence clinical decision-making, guidelines, and policy	Includes expert perspectives, critiques, and advancements in research methodologies
Studies included	Clinical trials, observational studies, diagnostic studies, longitudinal studies	Systematic reviews, meta-analysis, and narrative reviews	Case reports, case series, clinical practice guidelines, and health policy papers	Editorials, position papers, theoretical papers, methodology papers, and research protocols
Key focus of the abstract	New data and findings	Summarizing and interpreting existing studies	Practical implications for clinicians and healthcare providers	Expert insights, critiques, or new theories
Guidelines for writing	CONSORT for randomized control trials (guidelines for abstract) STROBE for observations studies	PRISMA 2020 for systematic reviews and meta-analysis	CARE for case reports and series AGREE for clinical practice guidelines	

Table 6.5 Key considerations for abstract introduction

	Original research	Review articles	Clinical application and practice	Commentary, opinions, and methodological advances
Introduction	Provide brief overview of the **background** and the purpose Concisely state the **research question** or **hypothesis**	Clearly state the **research question** that the synthesis addresses Explain the **importance** of synthesizing these studies [14]	Establish **clinical context** to the clinical case and the **rationale** for the case series What is **unique** about the clinical issue and what is its **significance**? [15]	Introduce the **topic or theory** being discussed State why this issue is important and what **gap it addresses**

The introduction answers the *"why"* question of your research, i.e., why is your study important?

1. The abstract should **identify the background**, i.e., what is already known about the subject to establish context and the knowledge gap or the research problem [11].
2. It should outline what your study intends to examine: write the **primary goals** and **objectives** and explain the **hypothesis** [6, 8].

Common Error
Too Much Background Information
- Problem: Overloading the abstract with background information rather than focusing on the study itself [11].
- Example: "Obesity has been a growing problem worldwide, with many health risks associated with it. Various diets have been proposed to tackle obesity. Among these, the low-carbohydrate diet has been particularly popular. This study looks at the effects of a low-carbohydrate diet."
- Solution: Keep the background information brief and relevant.
- Example: "Obesity is a growing problem worldwide. This study examines the effects of a low-carbohydrate diet on weight loss in obese adults."

6.2.8.2 Methodology
The methodology section answers the "How" of the research, giving a snapshot into how your study was conducted.

Common Error
Vague Methodology:
  - Problem: Failing to mention the details that are important and using generic quantification [11].
  - Example: "We reviewed the literature on many search engines (e.g., PubMed, Cochrane Library, etc.) to evaluate the effectiveness of the treatment of drug X in the treatment of coronary artery disease."

Table 6.6 Key considerations for writing abstract methodology

	Original research	Review articles	Clinical application and practice	Commentary, opinions, and methodological advances
Methodology	Provide details on the **study design** (e.g., randomized controlled trial, cohort study, diagnostic study), **sample size, setting, eligibility criteria, interventions, analytical methods,** and **key variables** measured [13]	Describe the **information sources, eligibility criteria,** methods to assess the **risk of bias,** and the **synthesis of results** [14] For a meta-analysis, mention any **statistical methods** used to combine study results [14] Narrative reviews offer a more flexible approach to synthesizing the literature	For case reports, describe the **case presentation**: The **history, investigations,** and main **clinical findings** [15] If developing guidelines, describe the **methodology** used	If the article is a **theoretical paper** or **methodological paper**, briefly describe the **new approach, model,** or **framework** being proposed If it is a commentary, outline the **critiques** of the existing literature or study

- Solution: Adding all the details that are necessary.
- Example: "We conducted a systematic review and meta-analysis following PRISMA guidelines. We searched PubMed, Cochrane Library, and EMBASE for studies published between 2015 and 2024 that assessed the effectiveness of the drug X in the treatment of coronary heart disease. Two independent reviewers screened titles and abstracts and extracted the data. Meta-analysis was performed using the random effects model to calculate pooled effect sizes and 95% confidence intervals."

6.2.8.3 Results
- This section is arguably the most important section of your abstract as most readers are interested in the findings of the study.
- Results should be presented in a logical order: chronologically and according to importance. (For e.g., experimental results come first then followed by control, details of the results are presented in the same sequence as the outlined in the study question, changed variables come before unchanged ones, etc.)
- The results should be written in the past tense and be confined to the study's scope [7].

Table 6.7 Key considerations for abstract results

	Original research	Review articles	Clinical application and practice	Commentary, opinions, and methodological advances
Results	Summarize the **primary outcomes** (quantitative results if possible, e.g., mean differences, relative risks, and p-values) [13] Mention any **significant findings** and the **effect size** if relevant Be specific about the results without overloading with numbers	Mention the number of **included studies** [14] Present the **synthesis of results** Including any **statistical results in meta-analysis** (e.g., pooled effect size and confidence intervals) [14] If the results are narrative, summarize the **overall trends**	Present **key findings** For case reports, describe the **diagnosis/findings, interventions/treatments**, and **clinical outcomes** [15] The difference between the results and methodology section for case reports is not clear cut, and both sections are usually written together	Present the **key insights**, Findings, or recommendations derived from the theory, critique, or methodological proposal

Common Errors
1. **Inconsistent or Incomplete Results** [6]
    Problem: Providing incomplete or inconsistent results or focusing only on significant findings.
    – Example: Our findings suggest some improvement in cognitive function.
    Solution: Summarize all major findings and ensure consistency with the results presented in the full paper.
    – Example: "Our findings suggest a 25% improvement in cognitive function scores among participants who took the supplement for 15 weeks."
2. **Focusing on Secondary Results**
    Problem: When the primary results addressing the primary objective are negative or lack statistical significance, authors highlight the secondary results.
    – Example: Although the primary outcomes of the study did not show statistically significant differences in the rates of complications between the two treatment groups ($p = 0.45$), we observed a significant reduction in hospital stay duration in the secondary analysis, with a mean difference of 2.5 days ($p = 0.02$).

Solution: Authors should give prominence to the primary results even if they are negative. Secondary results should get only a single brief sentence.
    – Example: "The primary outcomes of the study indicated no statistically significant differences in the rates of complications between the two treatment

Table 6.8 Key considerations for abstract conclusions

	Original research	Review articles	Clinical application and practice	Commentary, Opinions, and methodological advances
Overview	Highlight the **implications** of the results and their relevance to healthcare practice For clinical trials, mention the **funding source** and **trial registration number** [13]	Summarize the **implications** of your findings If the findings suggest clear conclusions, state them Mention the **limitations** of evidence [14]	Emphasize the **clinical implications of the case** and any **recommendations** if they emerged What is the main "take-away" lesson from the case? [15]	Provide **recommendations** for future research, practical applications, or policy changes based on the insights or critiques

groups ($p = 0.45$). However, secondary results showed a noteworthy reduction in hospital stay duration, with a mean difference of 2.5 days ($p = 0.02$), suggesting a potential area for further investigation."

6.2.8.4 Conclusion

The conclusion section of an abstract addresses the question, "so what?" by **interpreting the findings and discussing their broader implications** in the field.

1. Consider whether the results can be applied to other contexts, if they address the knowledge gap identified in the introduction, how they compare to related studies, and if they suggest new hypotheses [11].
2. To maintain professionalism and accuracy, avoid overgeneralizing or exaggerating the implications; focus strictly on the data presented and logically connect the key findings.

Common Error
1. **No Clear Conclusion**
   - Problem: Failing to provide a clear conclusion or the implications of the research.
   - Example: "The study's findings are discussed."
   - Solution: Summarize the main conclusions and highlight the significance of the findings.
   - Example: "The study concludes that a high-fiber diet significantly reduces the risk of colon cancer."

6.2.9 General Qualities of a Good Abstract (Fig. 6.4) [6, 11]

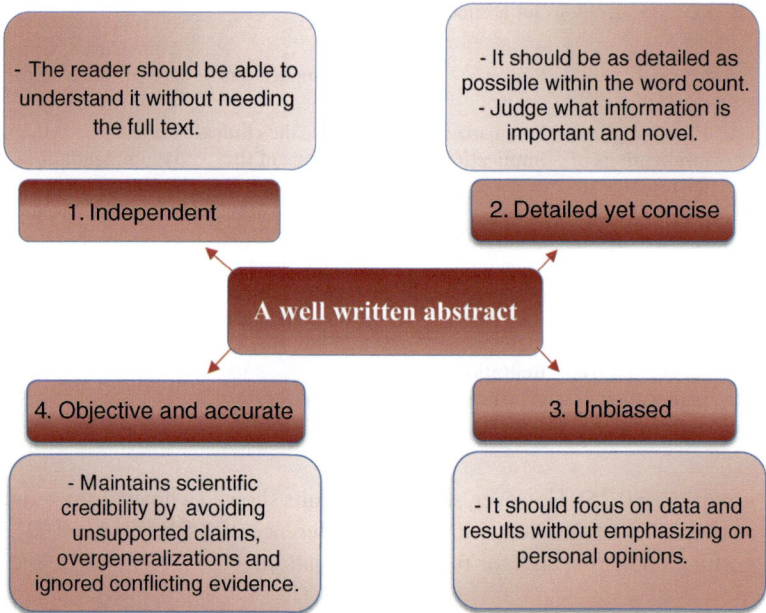

Fig. 6.4 A well-written abstract

6.2.10 Don'ts

1. Do not exceed the word limit.
2. Do not add tables and figures [16].
3. Do not cite references.
4. Do not use acronyms and abbreviations unless the term appears multiple times, and explain them first.
5. Do not include information that is not present in the paper.
6. Avoid using unnecessary jargon: It makes the paper unnecessarily technical and may not be understood by a broader audience.

6.2.11 Reviewing an Example of the Abstract

– **Meta-Analysis** [17]

"Purpose of Review: The combination of ketamine with propofol and dexmedetomidine has gained popularity for sedation and general anesthesia in different populations. In our meta-analysis, we helped the anesthesiologists to know the efficiency and the efficacy of both combinations in adult and pediatric patients.

Introduction:
- Context is given to the article's significance in the field.
- Clear objective: the goal is to provide insight into the efficacy of these combinations in adult and pediatric populations.

Methods: We searched PubMed, CENTRAL, Web of Science, and Scopus from inception to August 1, 2023. Our outcome parameters for efficacy were recovery time, pain score, and physician satisfaction while for safety were the related cardiorespiratory, neurological, and gastrointestinal adverse events.

Methodology:
- The comprehensive search strategy has been mentioned.
- Defined timeframe: Allows readers to determine relevancy.
- clinically relevant outcome parameters.

Recent Findings: Twenty-two trials were included with a total of 1429 patients. We found a significantly longer recovery time in the ketadex group of 7.59 min (95% CI, 4.92, 10.26; I2 = 94%) and a significantly less pain score of − 0.72 (95% CI, − 1.10, − 0.34; I2 = 0%). Adults had a significantly better physician satisfaction score with the ketofol group, odds ratio of 0.29 (95% CI, 0.12, 0.71; I2 = 0%). Recovery agitations were higher in the ketofol group with an odds ratio of 0.48 (95% CI, 0.24, 0.98; I2 = 36%). Furthermore, we found a significant difference between the combinations with a higher incidence in the ketadex group with pooled odds ratio of 1.75 (95% CI, 1.06, 2.88; I2 = 15%).

Results:
- Clear and precise: It presents numeric findings.
- Confidence intervals and I2 values demonstrate statistical rigor
- Effective odds ratios: This compares outcomes making interpretation easier.
- Information is concise and ordered logically.

Summary: Ketadex was associated with lower pain scores, hypoxic events and airway obstruction, and emergence agitation. At the same time, ketofol had much more clinician satisfaction which might be attributed to the shorter recovery time and lower incidence of nausea and vomiting. Therefore, we suppose that ketadex is the better combination in periprocedural sedation for both adult and pediatric patients who are not at greater risk for postoperative nausea and vomiting."

Conclusion:
- Effective summarization: Key findings are highlighted.
- Balanced comparison: It compares the combinations fairly.
- Practical implications: It offers an evidence-based recommendation.

- **Case Report** [18]

"Introduction In this study, two cases that demonstrate the importance of bedside echocardiography and hands-off telemedicine technology for diagnosis and intervention in patients with coronavirus disease 2019 (COVID-19) are discussed. Case Presentation We report two cases of cardiac emergency associated with COVID-19. Case 1 is a 50-year-old female patient with chronic hypertension and chronic renal failure. Case 2 is a 64-year-old female with atrial fibrillation and recent stroke. Both were admitted to an isolation intensive care unit that was designated specifically to patients with COVID-19. Conclusions During admission, both patients had sudden deterioration characterized by oxygen desaturation and hypotension necessitating inotropic support. As a result, for both patients, bedside echocardiography was performed by the attending intensivist. Echocardiographic findings showed cardiac tamponade and acute pulmonary embolism, respectively, which were confirmed by a cardiologist through telemedicine technology. Proper emergency management was initiated, and both patients recovered well. Limited bedside transthoracic echocardiography had a front-line impact on the treatment and outcome of the two patients with COVID-19. By implementing telemedicine technology, the lives of two patients were saved, demonstrating the significance of telemedicine in isolation intensive care units in the developing countries during the COVID-19 pandemic."	- Establishes clinical context and significance. - Shows how the case is unique by addressing resource-limited settings in developing countries. Case description is chronologically sequenced in order of: - History and presentation - Main issue - Intervention that led to the - Critical findings - Treatment - Clinical outcome - **Clear takeaway message**: -Shows the **life-saving potential** of bedside echocardiography and telemedicine. -Emphasizes their importance in **developing countries** and similar pandemic scenarios.

6.2.12 AI Tools

The following artificial intelligence tools can be utilized to refine and enhance the quality of your abstract.

1. **Jenni AI**

    Jenni AI can elevate your academic writing experience and accelerate your journey toward academic excellence.
    https://jenni.ai/for-researchers

2. **paperpal**

    Boost your chances of success with real-time, subject-specific language suggestions that help you write better, faster!
    https://paperpal.com/

3. **avidnote**

    https://avidnote.com/

4. **perplexity.ai**

    Your AI-powered research & collaboration hub customized for your purpose
    https://www.perplexity.ai/spaces

References

1. Faryadi Q. PhD thesis writing process: a systematic approach—how to write your methodology, results and conclusion. Creative Education. 2019;10:766–83.
2. Research Guides at Sacred Heart University, Organizing academic research papers: 9. The conclusion. *Sacred Heart University Library*. (cited 2024 November 5,) URL: https://library.sacredheart.edu/c.php?g=29803&p=185935
3. Bouchrika I. How to write a conclusion for a research paper: effective tips and strategies in 2024. Research. 2024; URL: https://research.com/research/how-to-write-a-conclusion-for-a-research-paper
4. Faryadi Q. How to write your PhD proposal: a step-by-step guide. Am Int J Contemp Res. 2012;2(4):111–5.
5. Mohammad Shehata I, Masood W, Nemr N, Anderson A, Bhusal K, Edinoff AN, Cornett EM, Kaye AM, Kaye AD. The possible application of ketamine in the treatment of depression in Alzheimer's disease. Neurol Int. 2022;14(2):310–21.
6. Andrade C. How to write a good abstract for a scientific paper or conference presentation. Indian J Psychiatry. 2011;53(2):172–5.
7. Bouchrika I. * How to write a research paper abstract in 2024: Guide With Examples. 2024. * Research.com
8. Riordan L. Mastering the art of abstracts. J Am Osteopath Assoc. 2015;115:41–5. https://doi.org/10.7556/jaoa.2015.006.
9. Hartley J. Current findings from research on structured abstracts: an update. J Med Libr Assoc. 2014;102(3):146–8.
10. Tohid H. Unstructured abstract in academic writing: a comprehensive guide. CIBNP. 2023. https://www.cibnp.com/unstructured-abstract-in-academic-writing-a-comprehensive-guide/.
11. Bahadoran Z, Mirmiran P, Kashfi K, Ghasemi A. The principles of biomedical scientific writing: abstract and keywords. Int J Endocrinol Metab. 2020;28:18.
12. Types of journal articles, Author & reviewer tutorials, Springer. Available from: https://www.springer.com/gp/authors-editors/authorandreviewertutorials/writing-a-journal-manuscript/types-of-journal-articles/10285504
13. Schulz KF, Altman DG, Moher D. CONSORT 2010 statement: updated guidelines for reporting parallel group randomised trials. PLOS Med. 2010;7(3) https://doi.org/10.1371/journal.pmed.1000251.
14. Page MJ, McKenzie JE, Bossuyt PM, et al. The PRISMA 2020 statement: an updated guideline for reporting systematic reviews. PLoS Med. 2021;18(3):e1003583. https://doi.org/10.1371/journal.pmed.1003583.
15. Gagnier JJ, Kienle G, Altman DG, et al. The CARE guidelines: Consensus-based clinical case report guideline development. J Clin Epidemiol. 2013;66(11) https://doi.org/10.1016/j.jclinepi.2013.05.004. Manuscript preparation: abstract. In: AMA Manual of Style: A Guide for Authors and Editors. 10th ed. New York, NY:Oxford University Press; 2007:20–24
16. American Psychological Association, issuing body, & American Psychological Association, issuing body. Publication manual of the American Psychological Association: the official guide to APA style. (Seventh edition.). American Psychological Association. 2020.
17. Elsaeidy AS, Ahmad AHM, Kohaf NA, Aboutaleb A, Kumar D, Elsaeidy KS, Mohamed OS, Kaye AD, Shehata IM. Efficacy and safety of ketamine-Dexmedetomidine versus ketamine-Propofol combination for Periprocedural sedation: a systematic review and meta-analysis. Curr Pain Headache Rep. 2024;28(4):211–27. https://doi.org/10.1007/s11916-023-01208-0. Epub 2024 Jan 12
18. Mohammed Sheata I, Smith SR, Kamel H, Varrassi G, Imani F, Dayani A, Myrcik D, Urits I, Viswanath O, Taha SS. Pulmonary embolism and cardiac tamponade in critical Care patients with COVID-19; telemedicine's role in developing countries: case reports and literature review. Anesth Pain Med. 2021;11(2):e113752. https://doi.org/10.5812/aapm.113752.

The Artwork

7

Ro'a Azzam, Islam Mohammad Shehata, Omar Viswanath, and Shaleen Vira

> *Ability to draw and communicate visually can no longer be seen as optional.*
>
> —Bette Fetter

7.1 Definition

Any graphical display used to present information or data.—The AMA Manual of Style

7.2 Importance of Artwork (See Fig. 7.1)

- Readers, peer reviewers, and editors reported that figures and tables are one of the first parts of the manuscript that they pay attention to [1].

R. Azzam (✉)
Faculty of Medicine, Modern University for Technology and Information, Cairo, Egypt
e-mail: Roaa.94719@medicine.mti.edu.eg

I. M. Shehata
Department of Anesthesiology, Faculty of Medicine, Ain Shams University Cairo, Cairo, Egypt
e-mail: Islam.shehata@med.asu.edu.eg

O. Viswanath
Department of Anesthesiology, Creighton University School of Medicine, Phoenix, AZ, USA

Mountain View Headache and Spine Institute, Phoenix, AZ, USA

S. Vira
Vice Chairman, Business Development Associate Professor, Orthopaedic Surgery & Neurosurgery Banner Health, University of Arizona School of Medicine, Phoenix, AZ, USA

© The Author(s), under exclusive license to Springer Nature Switzerland AG 2025
I. M. Shehata, O. Viswanath (eds.), *How to Successfully Publish a Manuscript*,
https://doi.org/10.1007/978-3-031-92538-2_7

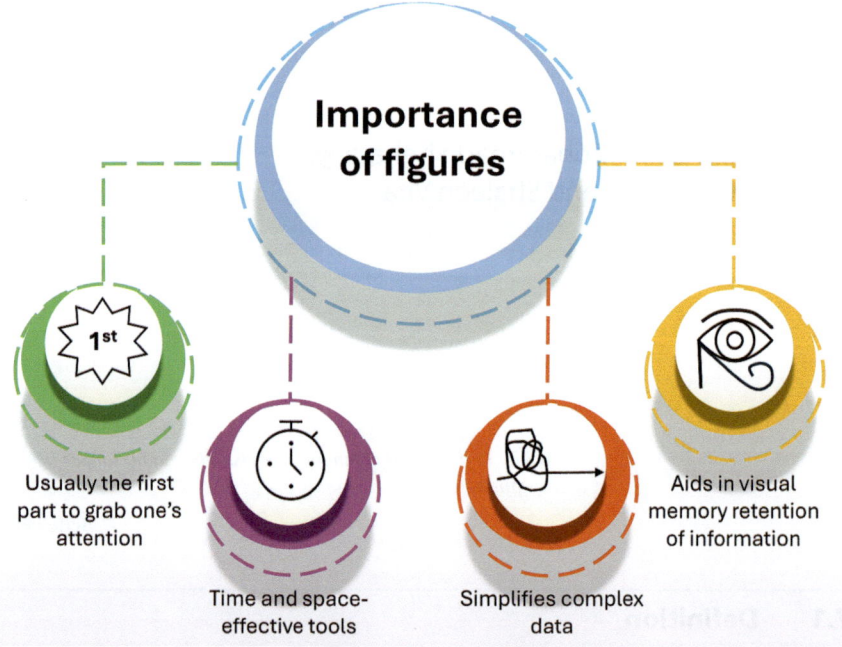

Fig. 7.1 Designed on Microsoft PowerPoint

- For that reason, journals started to include thumbnails of the figures in a paper below the abstract [1].
- A well-designed figure delivers a message clearly that is easy to recall.

7.3 Types of Figures (See Fig. 7.2)

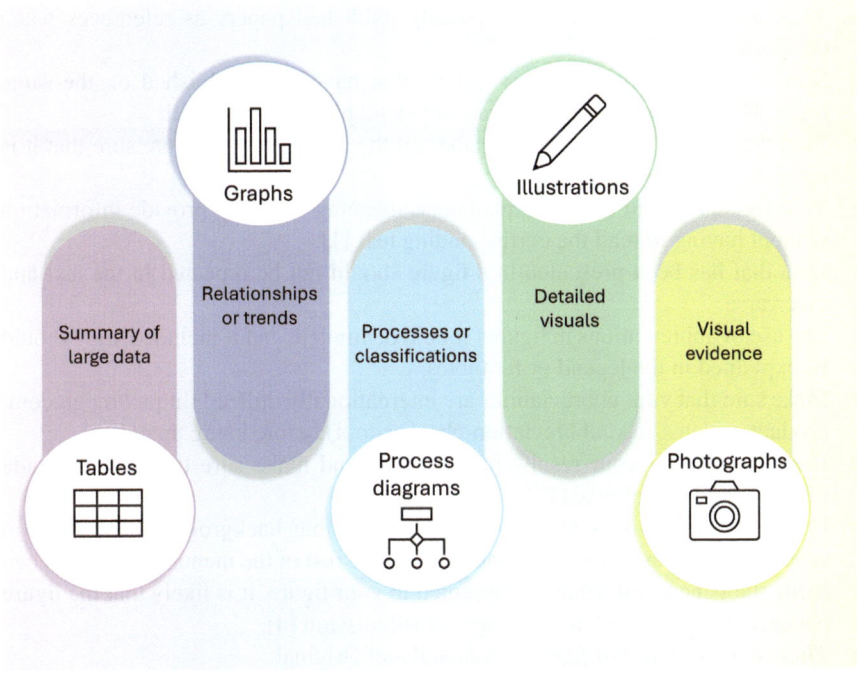

Fig. 7.2 Types of figures designed on Microsoft PowerPoint

7.4 General Considerations in Preparing Figures

- Always check the ***author guidelines*** for your target journal regarding the size, number, style of figures, as well as the use of AI in design.
- You can also use figures from already published papers as references when designing [2].
- Run a "literature review" over figures that have been published on the same topic. This will help you find ideas and avoid plagiarism as well.
- Start drafting a figure after you gather all the data needed and are sure that it is relevant and has an indication.
- Your figures should be self-explanatory, meaning that they provide information without having to read the corresponding text [1].
- Data that has been presented in a figure should not be repeated in the text and vice versa.
- The use of abbreviations in figures should be limited, and if included, they should be explained in the legend or footnotes.
- Make sure that your abbreviations are internationally utilized (https://megit.com/product-updates-feed/abbreviation-checker-tool) before using them [3].
- Double-check all your results for accuracy, and make sure that they coincide with the manuscript goals [2].
- Have your figures reviewed by readers from various backgrounds in an effort to ensure that they are understandable without the rest of the manuscript. If they can recite the general information presented in your figure, it is likely that the figure has met the benchmark for peer review submission [4].
- ***There are two types of figures***: Adapted and Original.

A. Adapted and Reprinted Figures.
- **Source:** Figshare.com is an online repository where you can find thousands of scientific figures available for use under different licenses.
- If the content is published under Creative Commons license (CC by 4.0), then you do not need permission, but proper attribution is required.
- However, if permission is warranted, **ask** for it and give credits for using any published figure and mention any alterations you have made over the original work [3].
- Before **asking**, know the exact number of the figure(s) you want to reuse and the type of use, title, publisher, lead author, format (print or electronic), and circulation of your own manuscript.
- After permission, add the figure reference in the legend with "adapted from" or "reprinted from" and in the references list as a regular citation.
- The copyright holder of a figure can be the author or the journal, so.
1. Author: Write an email to the author asking for permission to use and state what you need the figure for and any adjustments you would add.
2. Journal: You will find a link to the rights and permissions on its homepage or with the article/book of interest.

3. Copyright Clearance Center's Rights link https://marketplace.copyright.com/rs-ui-web/mp is a service where you can ask for permission or be guided to where you can (the publisher or the author).
B. Original Figures.
- Each figure should have an object identifier which is a number corresponding to the order in which they are placed in the text (e.g., "fig. 1, fig. 2, etc.).
  – If there is only one figure, then there is no need for a number (e.g., "see figure," "table below") [5].
- Titles should be included at the top of the figure to guide visualization. Scale bars, sample size, and abbreviations are better placed at the bottom right, [4] as shown in Fig. 7.3.
- Keep in mind that printing may alter some details in your figures.
  – Always use high resolution for your figures and make sure any markers such as arrows are of appropriate size.
  – Print your figures in their final size before submitting to ensure clarity.
- Legends or captions are brief explanations of the figure and are placed below it. They should not provide an explanation for your results or extensive details [8]. They are not always needed, and direct labeling is generally preferred [4].
- Footnotes are placed at the bottom of the figure, but can be placed beside it to save space, and in two columns if they cannot fit in one [6]. You can use (*, †, $, §, q, **, etc.) or use superscript lowercase letters (e.g., $^{a, b, c}$, etc.) [2].
- When to add foot notes?
  – Explain the abbreviations or measuring units, e.g., GAPDH: glyceraldehyde 3-phosphate dyhydrogenase;/hpf: per high power field; mEq/L: milliequivalents in a liter.

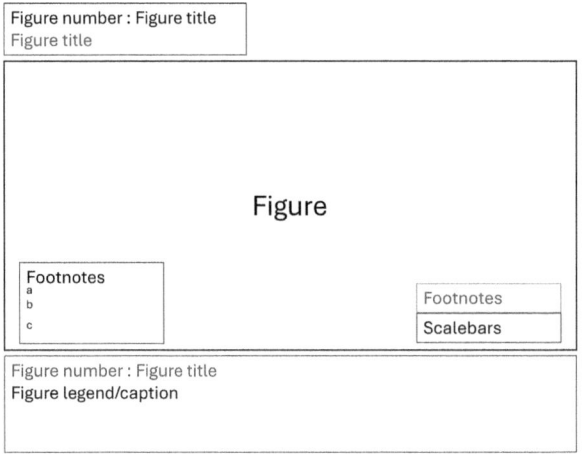

Fig. 7.3 Figure components. Gray typing indicates other placement options

- Refer to a section in your manuscript or supplementary data, e.g., see the Methodology section.
- Cite references or acknowledge any modifications done over an original figure, e.g., this figure has been reprinted from another source.
- Explain rounding of numerical data or scores, e.g., because of rounding, percentages may not total 100; NRS (numeric rating scale): 0 means "no pain" and 10 means "the worst pain imaginable."

7.5 Creating Tables

When?

- You have a large number of comparable or related data.

Why?

- They summarize and present the data concisely when alternatively explaining it in the manuscript would be challenging or less engaging.

Which program?

- SPSS.
- Microsoft Word for tables that have only text or more text than numbers. Microsoft Excel for tables that have numbers and calculations.
- Google Docs and Google Sheets allow you to share your work with other participants to view and edit online.

How?

- Generally, a table consists of an object identifier, a title, column and row headings, a body or data field, and footnotes if any [6]. See Table 7.1.
- Internal horizontal and vertical dividing lines in the field are not always necessary. Some journals suggest alternating shading of rows instead.
- Organize your table so that related data is read from above downward (in columns not in rows) [2].
- Left-justify your data in the table. Do not justify the margins within the cells in order not to impair readings [6].

Table 7.1 Title (content of a table)

Stub heading	Column heading (, unit)	Column heading (, unit)
Row heading	12	17
Row heading	15	13
Row heading	10	12

- To highlight a cell, you can type the entry in bold or italics or use a different color for shading and explain that in the footnotes [7].
- Include the measuring unit used next to your column headings instead of repeating it after each number, and add a comma before (e.g., mL;, mg/kg).
- If the data is repeated throughout a column, remove the column and include this entry in the footnotes, e.g., all studies are randomized clinical trials [6].
- When you have more than one table with the same headings, for example, tables that compare the same five groups at different times, keep the same order of the headings in all of the tables [9].
- Do not write "Group A" and "Group B" in your table; instead clarify what those groups are as "treated group" and "control group." This way, the table is clear without the text [9].
- Do not leave a cell empty if there is no data to type in. Instead, type N/A, Nil, or a simple dash "–."
- For long tables that cannot fit in one page, place the heading row at the top of each page of the table. For wide tables, use landscape orientation [7].
- Types of tables used in research papers are tables of study population characteristics, tables of results, tables of inclusion and exclusion criteria in systematic reviews and meta-analyses, data extraction tables, [10] and summary of results tables.
- Do not create a table with data that can easily be given in the text [9].

7.6 Different Types of Graphs

Why?

- Use graphs whenever you want to portray relationships, trends, and statistics. Table 7.2 shows a summary of different graph types and their uses.

Table 7.2 Types of graphs

Line graphs	Shows changes over time (short periods)	
Column graphs	Compares groups or categories or changes over a long period of time	

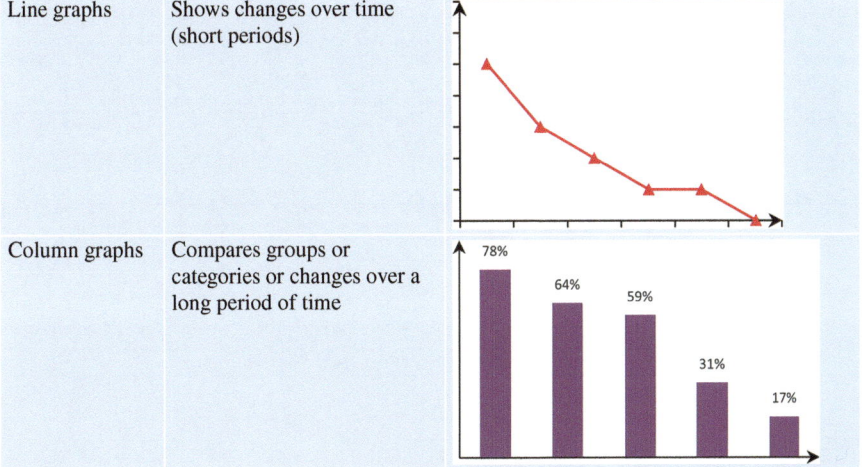

(continued)

Table 7.2 (continued)

Bar graphs	Compares large numbers of groups or categories or long column labels	
Histograms	Shows frequency distribution of data	
Scatterplots	Shows relationships or correlations between variables	
Pie charts	Shows percentages and proportions	

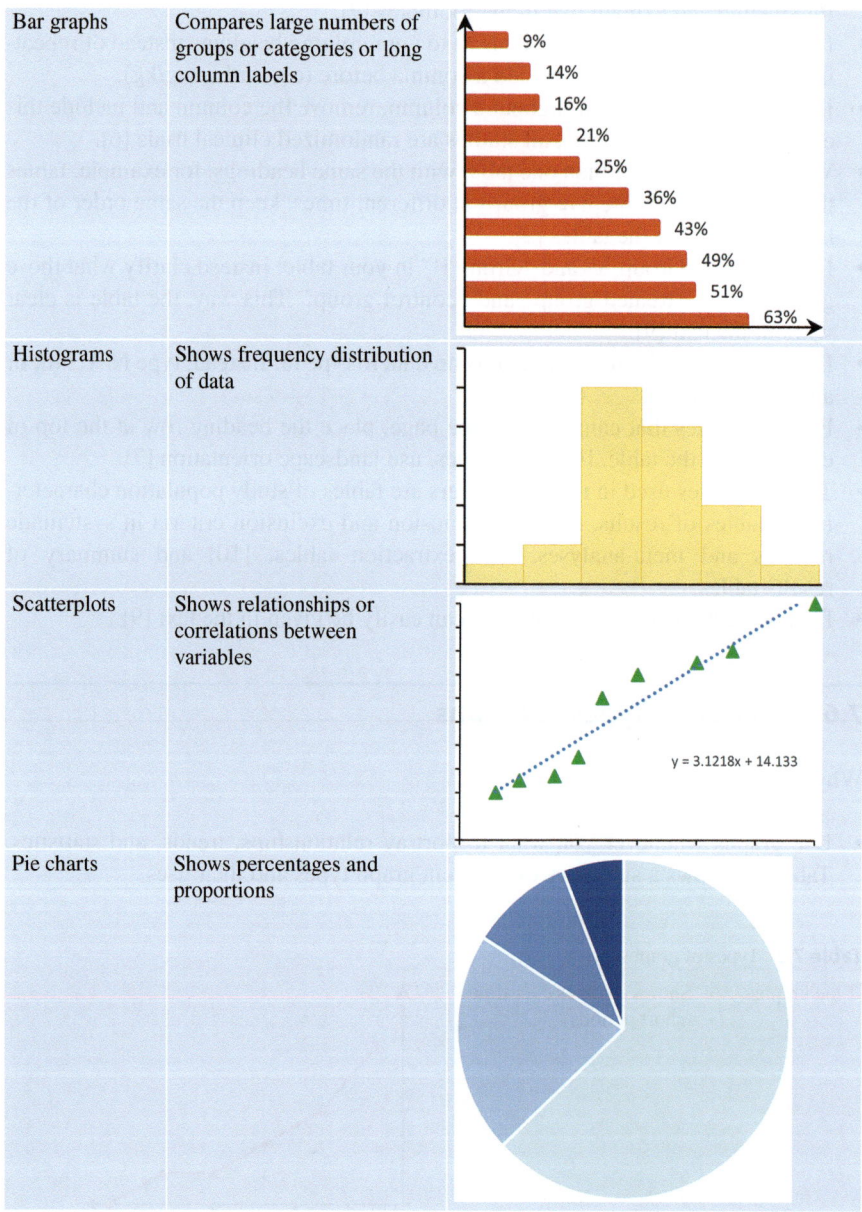

Which program?

- Microsoft Word, Excel, and PowerPoint can create all types of graphs.
- SPSS can create histograms, bar graphs, pie charts, line graphs, scatterplots, and box plots.
- Google Docs can create line graphs, pie charts, bar graphs, and column graphs.
- Google Sheets can create line graphs, pie charts, bar graphs, column graphs, histograms, area graphs, scatterplots, radar graphs, candlestick graphs, and waterfall graphs.
- Online tools as Piktochart, Livegap charts, Visme, Chartblocks, and Biorender.com.
- MokkupAI.com allows you to create graphs and dashboards.

How?

- Any scale in a graph should start at zero, especially line graphs and bar graphs. Otherwise, results may not be expressed properly. See Fig. 7.5d [11] Sometimes, the scale cannot be started at zero, as in Fig. 7.8a, b.
- Label each axis and provide the measuring unit if any. Axis scales are preferred to be multiplies of 2, 5, and 10 [6].
- If your data points are between 0 and 95, the highest number on your axes should be 100, and the axis scales are 0,20,40,60..., etc. Over-extending the axes may falsely show low results. The ratio between the two axes should be 1:1 or 3:2 [8]Include tick marks on your axes. They are preferred over gridlines, and only one of them should be in the graph [12].
- Keep the gridlines to a minimum or avoid using them if possible.
- Explain any arrows or highlights in the figure caption.
- Do not overuse colors, shading, or patterns. Choose colors with different values, as these will still be recognizable in black and white print.
- Never design bar graphs, column graphs, or pie charts in three dimensions. It is commonly misinterpreted and does not have any significance as in Fig. 7.5c.

7.6.1 Line Graphs

- Ideal for showing changes over a short period of time (trends) or displaying relationships between two or more variables [13].
- Keep the dependent variables on the Y-axis (vertical) and the independent variable on the X-axis (horizontal).
- You can present more than one dependent variable as in "Fig. 7.4b" but avoid overcrowding with more than four lines.
- Differentiate between the two variables by using continuous and dotted lines or different data points (e.g., ●, ■, ▲, △, □, ○). Explain what each pattern or symbol rep-

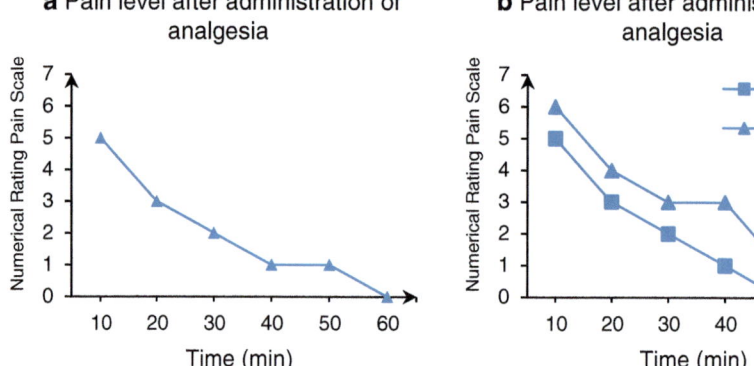

Fig. 7.4 Line graphs

resents in a Key or better label each line directly [14]. These symbols should be consistent throughout the manuscript. If you use circles for group X and triangles for group Y, then they should stay the same in all figures [11].

7.6.2 Column Graphs

- Design a column graph when you want to display categorical data. It can show changes over time, but each column is a set of data unrelated to a time series (unlike a line graph) [14].
- The width of the columns should all be equal and wider than the blank spaces between them.
- Always order your columns by size (ascending or descending) If possible.
- Try to stay away from patterned or black and white columns. Instead, color the columns in gray tones or colors with varying values.
- If there are more than one category in each group, it becomes a grouped column chart. Two to three categories are acceptable, but more than three in the chart will not be easily read, and this data is better presented in a table [14].
- Label the different categories in a grouped column either directly or in a Key. See Fig. 7.5b.
- Type the exact values of data over each column or bar if there are decimals.
- If your chart is overcrowded (column labels are overlapping or the chart has more than five columns), you may consider presenting the data in a bar chart [14].

7.6.3 Bar Graphs

- They are identical to column charts, but the bars (independent variables) here are shown horizontally. It is preferred if the data is too crowded in a column chart. See Fig. 7.6.
- They are also preferred if your data is not following a time or sequence.

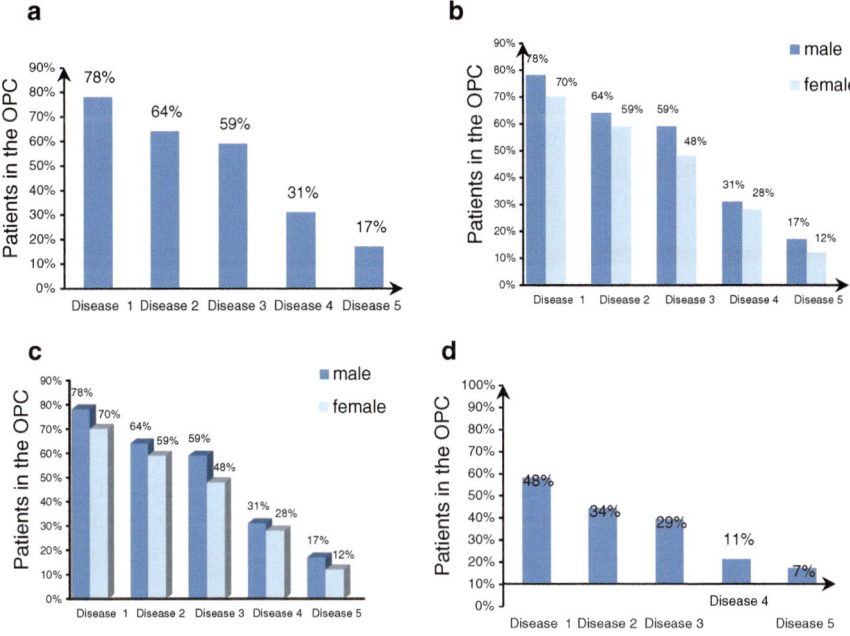

Fig. 7.5 Column graphs. *OPC* Outpatient clinic

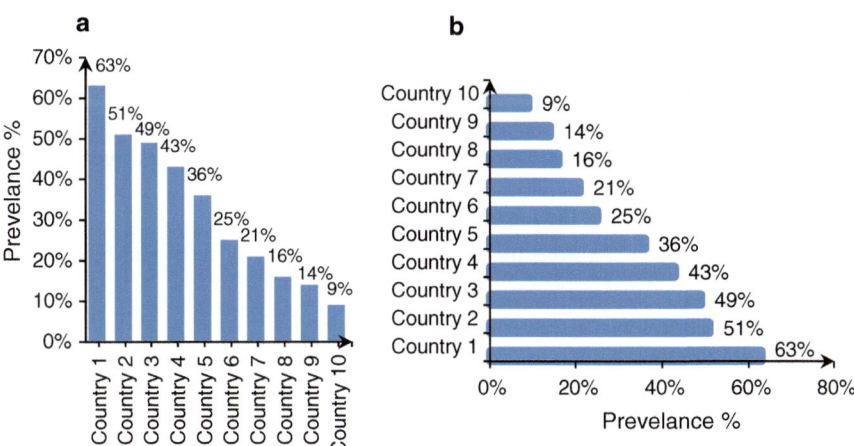

Fig. 7.6 Column graph (**a**) and bar graph (**b**) of prevalence of disease X among different populations

7.6.4 Histograms

- Design a histogram to show the distribution of continuous data along frequencies or percentages.
- The columns have equal widths and no gaps in between as shown in Fig. 7.7. This shows the continuity of data and differentiates it from column charts.
- 5 to 15 columns are ideal to display data correctly [15].

7.6.5 Scatter Plots

- They show relationships or correlations between two variables, especially how a change in the independent variable (X-axis) changes the dependent variable (Y-axis).
- Keep the data markers consistent in shape and color for each variable.
- Add a trendline to show the correlation between your variables. The relationship can be positive, negative, or null, as shown in Fig. 7.8a, b, c, respectively.
- Include the regression equation, P value, and sample size, either directly on the graph or in the caption [6].

7.6.6 Pie Charts

- They are not preferred in scientific publications and can be replaced by a table or a bar graph or even explained in written form in the manuscript [6].
- Begin the largest sector of the chart at 12 o'clock and then arrange the sectors in a descending order clockwise.
- Do not add more than five segments (replace the pie chart with a bar chart instead, as in Fig. 7.9c).
- If there is a segment that you want to highlight, you can detach it slightly from the chart.
- Place the labels outside the pie chart [14](Table 7.2)

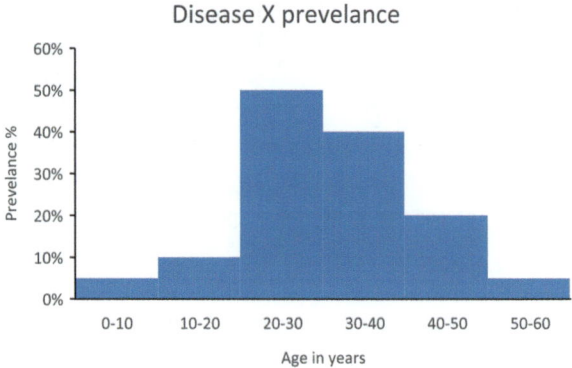

Fig. 7.7 Histogram

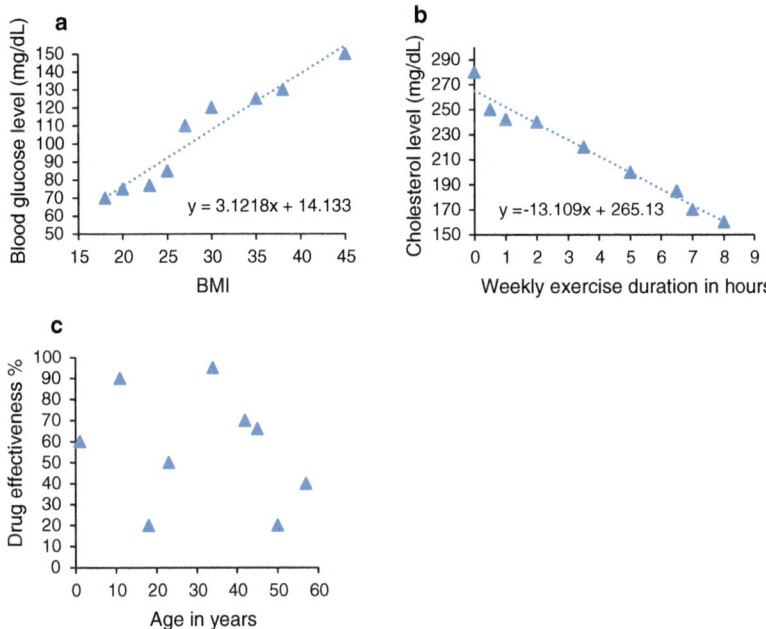

Fig. 7.8 Scatterplot (**a**) showing positive relationship between variables, (**b**) negative relationship, and (**c**) null

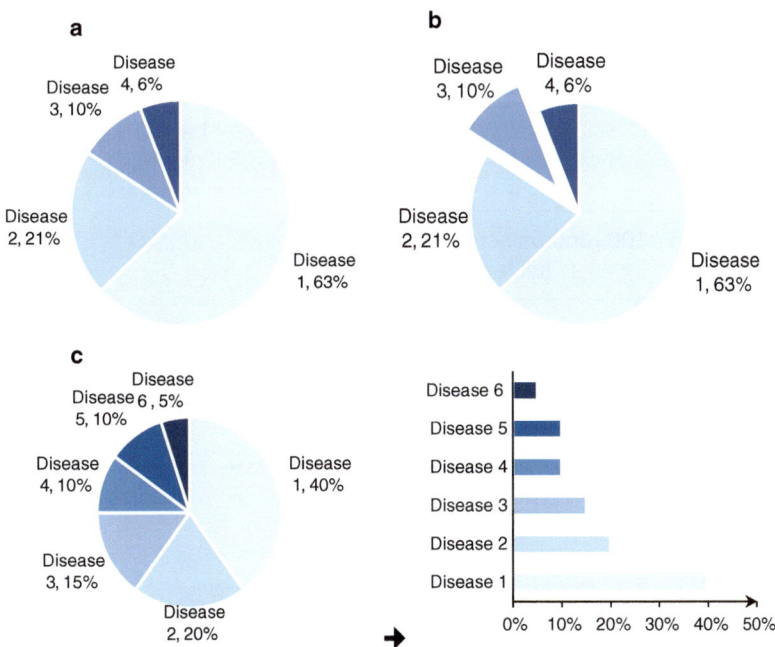

Fig. 7.9 (**a**) and (**b**) Different designs of piecharts. (**c**) Pie chart with too many segments better converted into a column chart

7.7 How to Make Process Diagrams?

Why?
- Process diagrams are used to present information in a symbolic, schematic, and simplified manner. Utilize them when you want to show a process, step-by-step instructions, decision-making, or classifications.
- Design a flowchart to display a study protocol, inclusion and exclusion of samples (PRISMA, Preferred Reporting Items for Systematic Reviews and Meta-Analyses) (see Fig. 7.10), recruitment, and follow-up of participants in clinical trials (CONSORT, Consolidated Standards of Reporting Trials) [6].

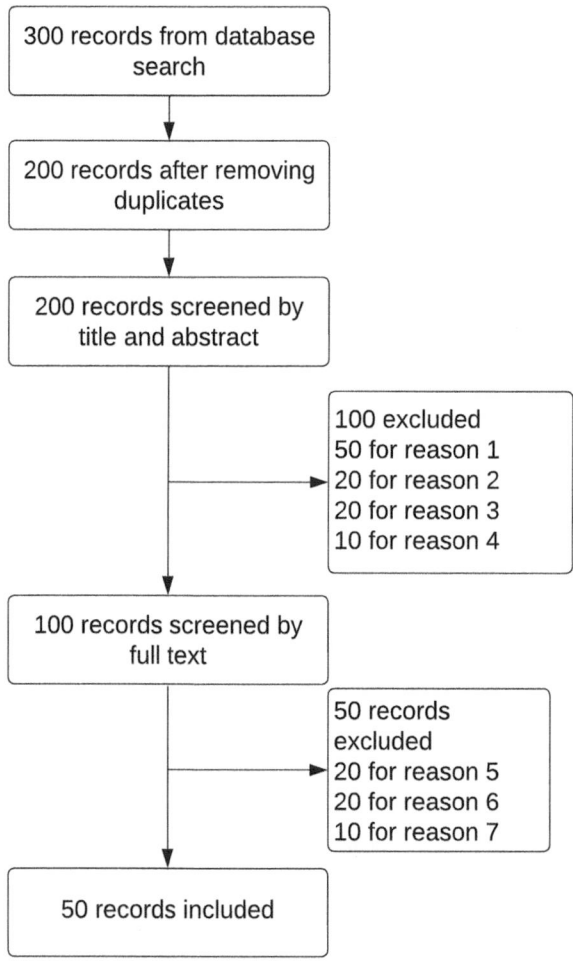

Fig. 7.10 PRISMA chart designed on Lucidchart.com

What Are They?
- Flowcharts and algorithms.

Which Programs?
- Microsoft Word and PowerPoint. The insert shapes or SmartArt option can easily help you with process diagrams.
- Online tools like Lucidchart are great alternatives.

How?
- Write down your process first and make sure it is simplified and properly ordered in sequence.
- Each step is placed in a box, and the flow between them is shown by lines or arrows.
- In a flowchart for a randomized clinical trial, the allocation point is placed in an oval box, while for an unrandomized trial, this step is placed in a rectangular box [6].
- CONSORT flowchart is a standardized template for reporting clinical trials shown in Fig. 7.11. It should include the enrollment, allocation, follow-up, and analysis steps.

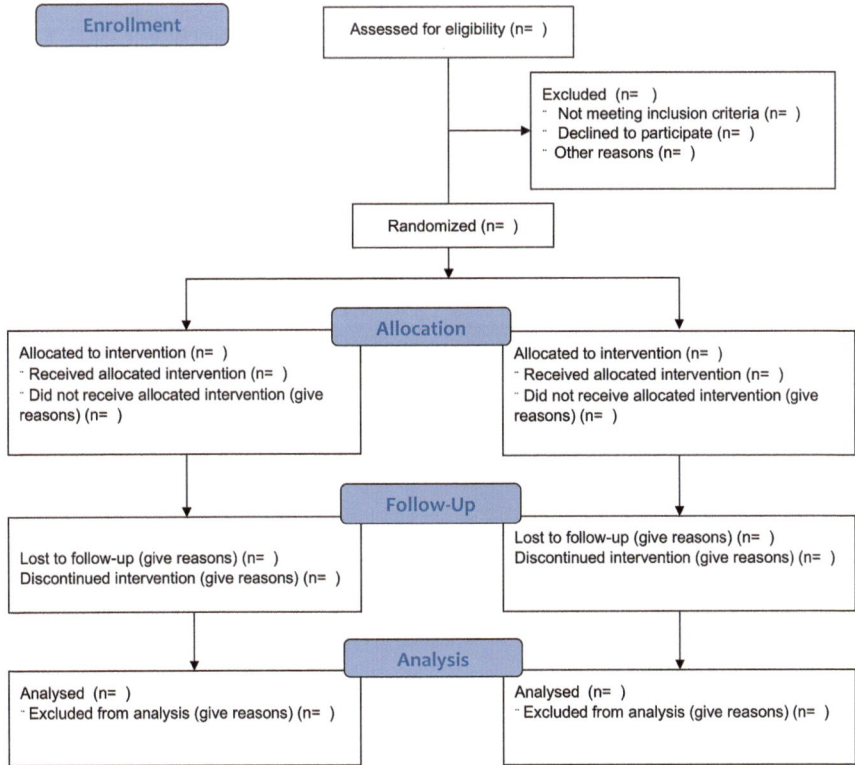

Fig. 7.11 CONSORT flowchart template

- In diagnostic algorithms, a decision box can be a hexagon or a rhombus and followed by the different possibilities (yes and no), [6] as in Fig. 7.12a. The entire diagram can be only rectangles and still be understandable as shown in Fig. 7.12b.
- Pedigrees are diagrams that show patterns of inheritance and have standard shapes for sex, vital status, carriers, and diseased.

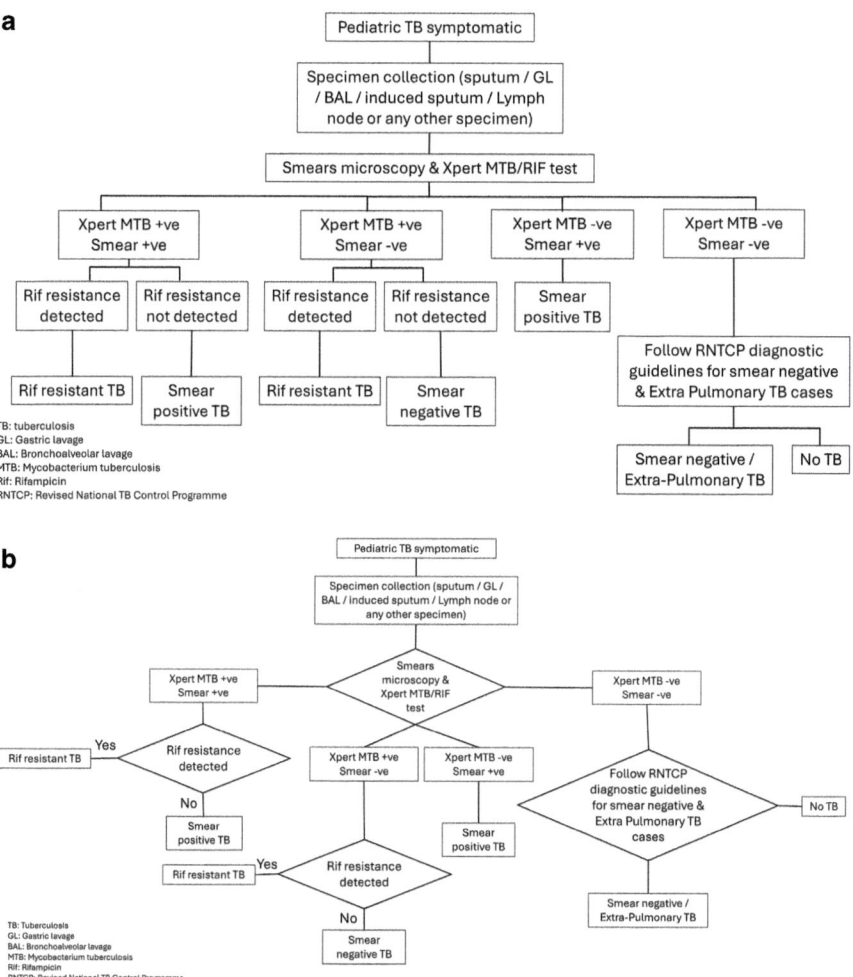

Fig. 7.12 (a) Diagnostic algorithm recreated on PowerPoint and abbreviations explained in footnotes. (Adapted from Raizada N. et al [16] under CC by 4.0 license). (b) Diagnostic algorithm recreated on PowerPoint and abbreviations explained in footnotes. (Adapted from Raizada N. et al [16] under CC by 4.0 license)

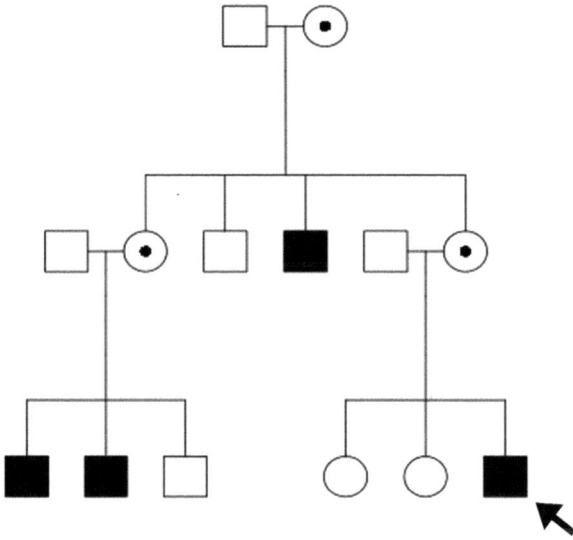

Fig. 7.13 Pedigree designed on Quickped.com [17]

- Use gender-neutral shapes in pedigrees if there is any problem with confidentiality or disclosing the gender is irrelevant as in autosomal traits, and explain that in the caption.
- Point to the proband with an arrow.
- Quickped is an online tool that creates pedigrees easily (see Fig. 7.13).

7.8 How to Design Illustrations?

What?
- They are visual representations of a concept or idea. They add an artistic touch and can be done in many styles.

Why?
- Design illustrations when you want to show anatomy, physiological pathways, and surgical procedures.

How?

- Hire a graphic designer or medical illustrator. You can design your own illustrations if you have an artistic background.

Programs?

Tool	Use
Adobe Photoshop, Adobe Illustrator	Excellent tools for designing
Biorender.com	Online scientific image and illustration software. It is easy to use and gives professional results. See Figs. 7.15 and 7.17
OnlineCamScanner.com	Scans your pictures and enables editing
Freepik, Visme.com	Online tools that supply thousands of high-quality images, icons, vectors, and templates. They also provide a designer tool that can help you design your illustration online
Venngage.com	Provides templates and AI-powered editing tools
Infogram.com	AI-powered tool that converts your text into infographics
Whimsical AI.com	Helps you create graphs, flowcharts, and dashboards and share them with your team to edit
Microsoft Powerpoint	For simple illustrations and infographics as Figs. 7.1 and 7.2
MolView, Chemdraw, ChemSketch, chemical sketch tool of RSCB	Great tools for illustrating chemical structures as in Fig. 7.19

- The first part you want to show in your figure should be placed in the top left corner, and the rest should progress to the bottom right corner. Alternatively, place your elements from top to bottom [18].
- Keep in mind "The Principle of Relevance" which states "Communication is most effective when neither too much nor too little information is presented." [18]
- To emphasize an element within your figure, use the design principal "Contrast." You can change the color, shape, size, or texture of an element to highlight it. Usually, one type of contrast is adequate [18].
- Do not overcrowd your illustrations with complex text [3] or too many colors. Figure 7.14a is an example of an overcrowded illustration.
- Do not miss out important details that should be included. Try to find another style for your illustration or remove any other irrelevant information instead. For example, change your illustration style from realistic to line art (Fig. 7.15, 7.16, 7.17, 7.18, and 7.19).

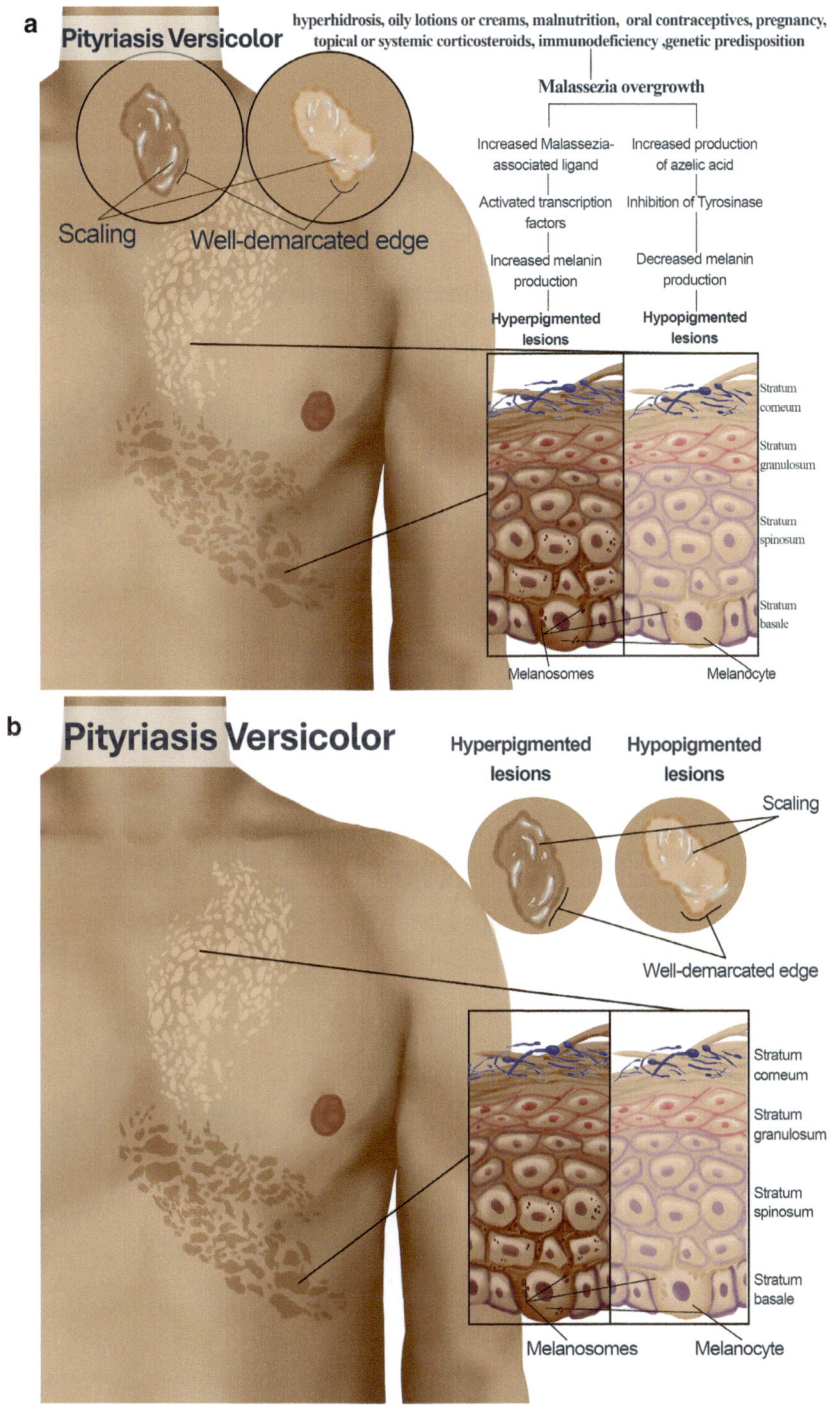

Fig. 7.14 (a) Overcrowded illustration with different font types and sizes. Designed on Photoshop. (b) Well-organized illustration with consistent font type and few text. Designed on Photoshop

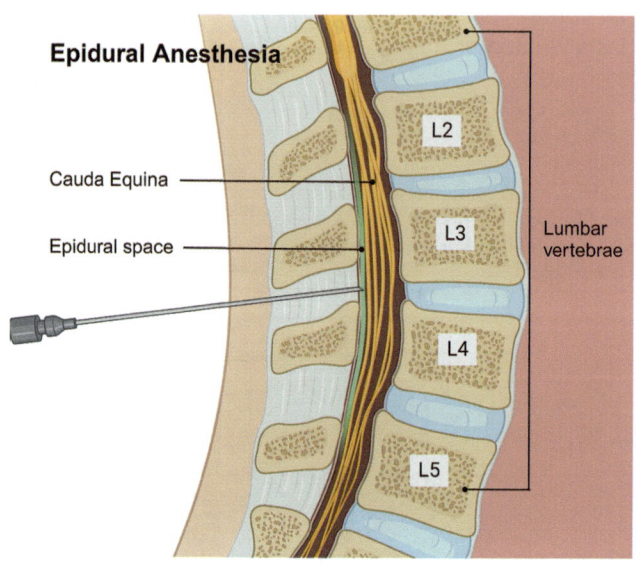

Fig. 7.15 Epidural anesthesia technique designed on BioRender.com. (Created in BioRender. ElSayed, M. (2025) https://BioRender.com/c42i395)

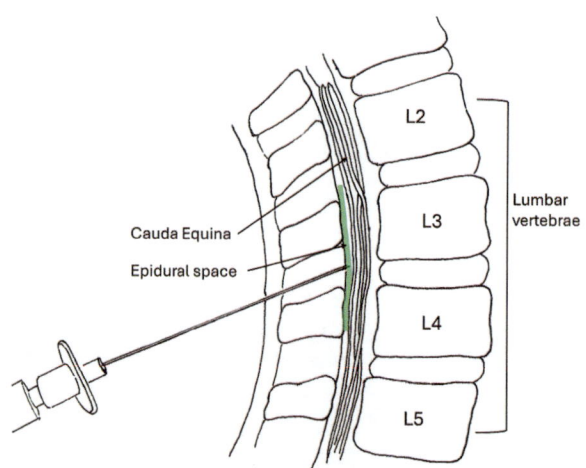

Fig. 7.16 Epidural anesthesia technique hand-drawn and scanned

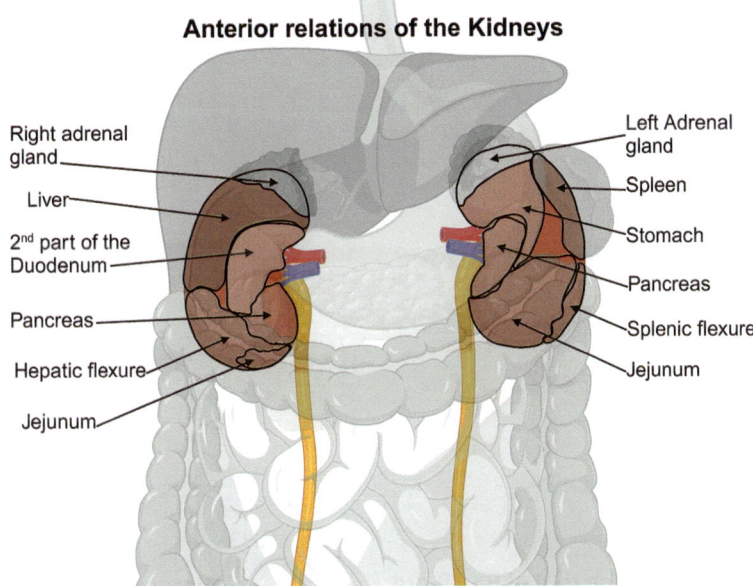

Fig. 7.17 Relations of the kidney designed on BioRender.com. (Created in BioRender. ElSayed, M. (2025) https://BioRender.com/w43e699)

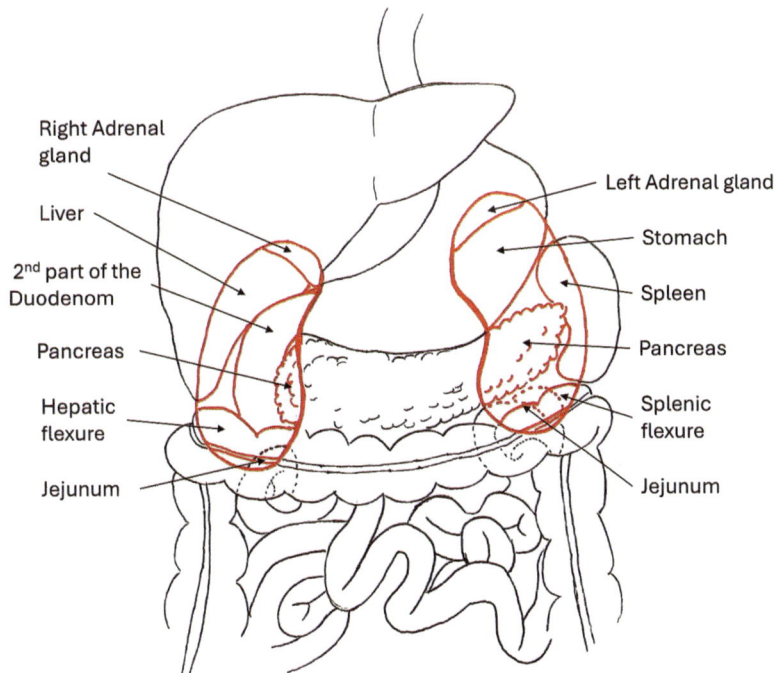

Fig. 7.18 Relations of the kidney hand-drawn and scanned

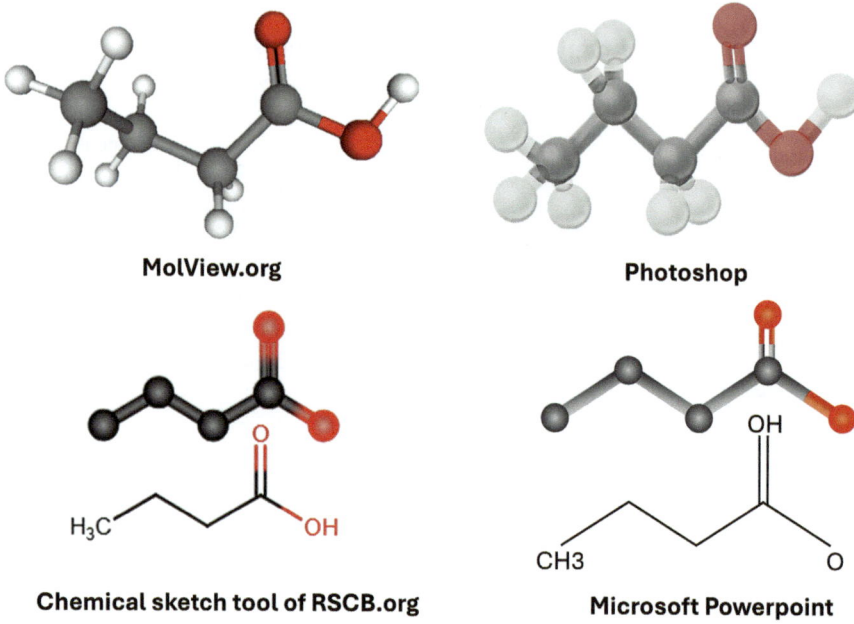

Fig. 7.19 Different ways to design chemical structures

7.9 How to Use Photographs?

Why?
- Visual evidence, document clinical findings such as "before and after" pictures [11]; steps in a surgical procedure.

Types?
- Photos of patients or specimens, radiographs, and microscopic photos.

How?

7.9.1 Photographs

- Take the photos with high resolution and good lightening.
- Keep the background clean and preferably of a solid color [11].
- Place the object of interest (specimen, patient, and body part) in the center of the frame and crop out any unnecessary parts [11].

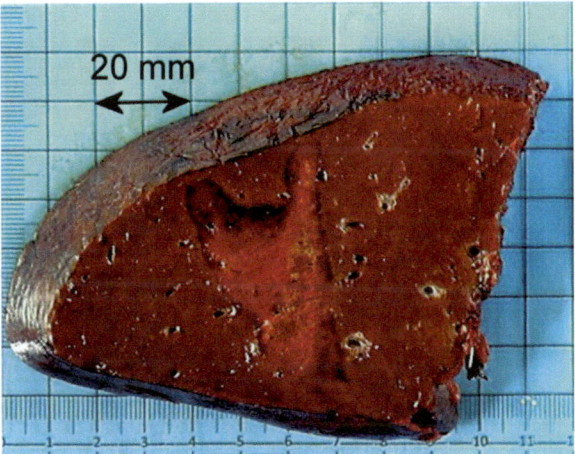

Fig. 7.20 Photograph of a specimen. (Reprinted from Dupré A et al [19] under CC by 4.0 license)

- Provide a scale bar or an object of a known size (e.g., a coin) next to the specimen. Figure 7.20 has a grid mat as a background, a ruler, and a scale bar. Any of these options should be sufficient.
- An informed consent should be obtained from the patient or the guardians if under 18 for using photos of identifiable parts of patients. Some journals do not require consent for photos of unidentifiable parts [20].
- Cover the eyes of the patient if they are not the subject of the photograph.

7.9.2 Radiological Images

- Describe the imaging modality and view of the image (e.g., coronal, sagittal etc.) [11].
- Label which side is left and which is right.
- Clarify the markers (arrows, arrowheads, etc.) and the dimensions of any scale bars either on the radiograph or in the legend, as in Fig. 7.21.

7.9.3 Microscopic Images

- Include a scale bar on the picture and describe the magnification power in the caption.
- Write the type of stain used in the caption. Figure 7.22 is a great example of how your picture and legend should look like.

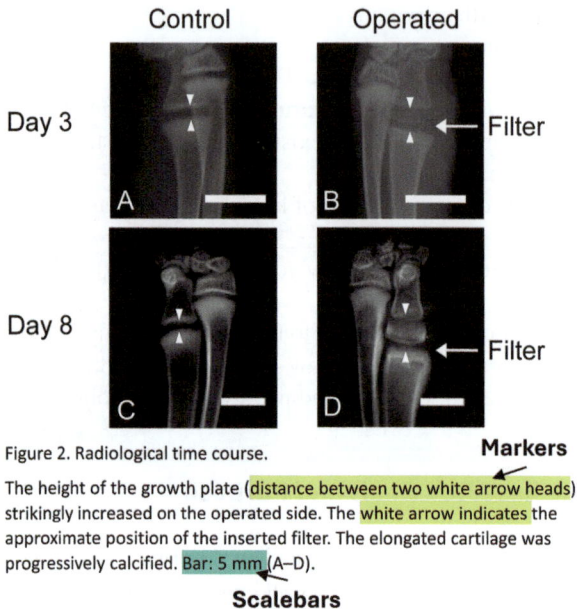

Figure 2. Radiological time course.
The height of the growth plate (distance between two white arrow heads) strikingly increased on the operated side. The white arrow indicates the approximate position of the inserted filter. The elongated cartilage was progressively calcified. Bar: 5 mm (A–D).

Fig. 7.21 Adapted from Enishi T et al [21] under CC by 4.0 license.

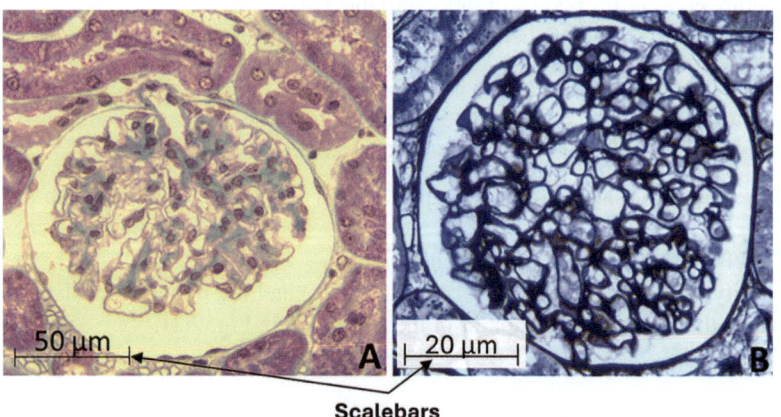

Fig 1. Light microscopy findings.
Control group: (A) Glomerular structure preserved, with no evidence of alterations. Masson's Trichrome (x400). (B) Preserved glomerular structure, without evidence of alterations. Jones methenamine silver (JMS) (x630).

Fig. 7.22 Adapted from 1. Crivellenti LZ et al [22] under CC by 4.0 license.

7.10 Submitting Your Figures

- **Color model:**
  - Create figures intended for digital display in RGB model, while those intended for print should be in CMYK model [23].
- **Image formats**

Text [23–25]

Image format	Advantage	Disadvantage	Use
.jpg or .jpeg	Small-sized file that is easy to upload	Can be pixelated due to compression	Digital images (nonprofessional)
.tif or .tiff	Details and colors are well conserved during transfer and editing	Large-sized file	Print and digital high-quality images
.png	Small-sized file and is not pixelated with compression	Cannot be saved as CMYK	Line diagrams and graphs
.eps,.ai,.pdf	Vector formats, so resolution is never lost	Large-sized files	Illustrations and diagrams
.raf,.dng,.nef,.srf,.heif	Colors and details are kept exactly as captured on camera	Large-sized files and need to be processed and converted to another format	Storing and editing photographs

- Make sure your figures fit the journal's guidelines regarding the dimensions, resolution, and file types before submitting.
- Check if your journal accepts ZIP (compressed) files. This is ideal for uploading multiple figures with high resolution.
- Journals have an online submission system to which you can upload your manuscript. They require you either embed the figures within the text or submit them in a separate file. You need to clearly indicate the placement of each figure, e.g., "insert fig. 1 here."
- Name each file as its object identifier in the manuscript, e.g.. "fig. 2.png, fig. 3.tiff".
- Add all permissions for adapted figures and consents for any photographs of human subjects.

7.11 Tips and Tricks

- Place your figure close to where it is mentioned in the manuscript [6].
- **Figure resolution:**
  - Keep your figures between 300 and 600 dpi. Line art and scanned graphics can be up to 1200 dpi.
  - Powerpoint default resolution is 96 dpi. You can change that from file → options → Advanced → Image size and quality → Default resolution. If you cannot find 300 dpi in the drop-down menu, follow this link to change the resolution settings. https://learn.microsoft.com/en-us/office/troubleshoot/powerpoint/change-export-slide-resolution
  - If your Powerpoint figure will be printed, set your image resolution to 300 dpi (or according to the journal guidelines). If it will be viewed digitally only, then set it to high fidelity.
  - Use an Online DPI Converter https://www.dpi-converter.com/ to change the resolution of any picture. This does not always improve the quality of the image, so try to start with a high resolution for all your figures.
- **Hand-drawn figures**
  - Use nontextured paper (regular printing paper) to achieve a smooth scan. Make sure that your pen strokes are sharp and thick as in figs. 16 and 18.
  - The photo you upload to an online scanner should be taken in adequate light, in focus, and high resolution.
  - If the quality of the photo is reduced, download your scanned figure as a PDF and then use an online file converter as Cloudconvert.com to adjust the quality and type of the final file.
- **Colors**.
  - Use a limited color palette [18] and try to keep it consistent throughout the figures to create homogeneity or a theme.
  - Your colors should be of different values to ensure that they are still readable in black and white [18].
  - Make sure that your figures are clear for people with visual impairments:
    - Color Oracle is a color blindness simulator https://colororacle.org/usage.html.
    - Combine magenta with green or cyan in highlighting or outlining. Adjust images that necessitate the presence of color such as fluorescence microscopy as Fig. 7.23 and mention the adjustments in the legend [1].
    - Avoid combining red and green together, and if it is not possible, then use them with varying values (brightness).You can find color blindness-friendly color combinations in this link:
    - https://cran.r-project.org/web/packages/colorBlindness/vignettes/colorBlindness.html#Collection_of_safe_colors
    - Color Contrast Analyzer https://www.tpgi.com/color-contrast-checker/ checks the contrast ratio in your figures and confirms if it passes the Web Content Accessibility Guidelines (WCAG).

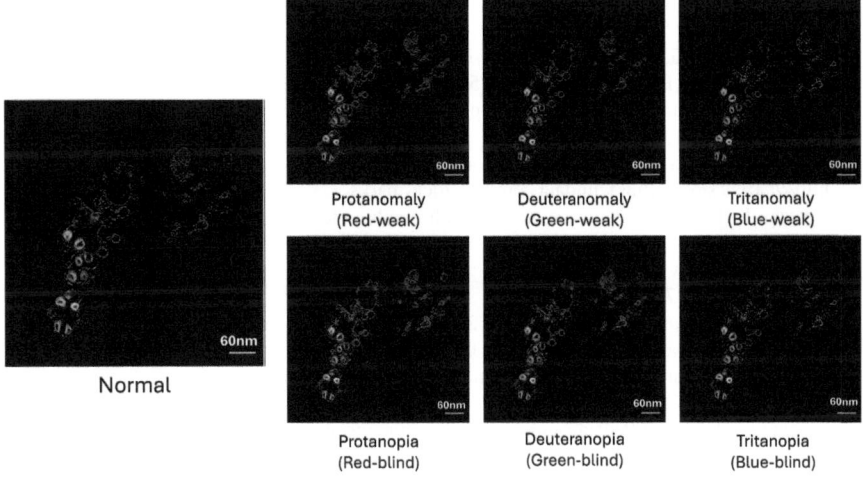

Fig. 7.23 Fluorescence microscopy picture, as seen by different color blindness types. (Adapted from Kunz TC et al [26] under CC by 4.0 license.)

- **Typing**:
  - Use sans serif fonts as **Arial, Calibri, Helvetica, Grotesque** [18].
  - Your font size should be between 8 and 14 points and consistent throughout the figure [7].
  - Embed your fonts in editable file types as .pdf, .ai, .docx, .eps.
  - Choose a font color that contrasts well with the background [18].
  - Do not place text over a patterned area in your figure [18]; instead, try placing your text in a colored box (see the scalebar in Fig. 7.22b) as long as no important details are lost or point with a labeled arrow if possible.
  - Do not overcrowd your figures with text. Label only the parts of interest.

References

1. Jambor H, Antonietti A, Alicea B, et al. Creating clear and informative image-based figures for scientific publications. PLoS Biol. 2021;19:e3001161.
2. Ng KH, Peh WC. Preparing effective tables. Routledge. 2002;eBooks:31–41.
3. Amobonye A, Lalung J, Mheta G, Pillai S. Writing a scientific review article: comprehensive insights for beginners. Sci World J. 2024;2024:1–13.
4. Jambor H. Better figures for natural sciences; 2018. https://doi.org/10.6084/m9.figshare.7001156.v2. https://pubmed.ncbi.nlm.nih.gov/19421674/
5. Ng KH, Peh WC. Preparing effective illustrations. Part 2: photographs, images and diagrams. Singapore Med J. 2009;50(4):330–4. quiz 335. PMID: 19421674
6. Committee NAM of S. AMA manual of style. Oxford University Press eBooks; 2020. https://doi.org/10.1093/jama/9780190246556.001.0001.
7. Publication manual of the American Psychological Association (7th ed.). American Psychological Association eBooks. 2019. https://doi.org/10.1037/0000165-000
8. Ng KH, Peh WCG. Preparing effective illustrations. Part 1: graphs. PubMed. 2009;50:245–9.

9. Gastel B, Day RA. How to write and publish a scientific paper. ABC-CLIO eBooks. 2022; https://doi.org/10.5040/9798400666933.
10. Inskip H, Ntani G, Westbury L, Di Gravio C, D'Angelo S, Parsons C, Baird J. Getting started with tables. Arch Public Health. 2017;75:14. https://doi.org/10.1186/s13690-017-0180-1.
11. Riordan L. Enhancing your manuscript with graphic elements, part 2: figures. PubMed. 2013;113:424–31.
12. Velez A. My plea for tick marks — storytelling with data. In: Storytelling with data; 2024. https://www.storytellingwithdata.com/blog/tick-marks.
13. Riordan L. Enhancing your manuscript with graphic elements, part 1: tables. J Osteopathic Med. 2013;113(1):54–7. https://doi.org/10.7556/jaoa.2013.113.1.54.
14. Gustavii B. How to write and illustrate scientific papers; 2008. https://doi.org/10.1017/cbo9780511808272.
15. Freeman JV, Walters SJ, Campbell MJ (2008) How to display data.
16. Raizada N, Khaparde SD, Salhotra VS, et al. Accelerating access to quality TB care for pediatric TB cases through better diagnostic strategy in four major cities of India. PLoS One. 2018;13:e0193194.
17. Vigeland MD. QuickPed: an online tool for drawing pedigrees and analysing relatedness. BMC Bioinformatics. 2022;23:220. https://doi.org/10.1186/s12859-022-04759-y.
18. Rolandi M, Cheng K, Pérez-Kriz S. A brief guide to designing effective figures for the scientific paper. Adv Mater. 2011;23:4343–6.
19. Dupré A, Melodelima D, Pérol D, Chen Y, Vincenot J, Chapelon J-Y, Rivoire M. First clinical experience of intra-operative high intensity focused ultrasound in patients with colorectal liver metastases: a phase I-IIa study. PLoS One. 2015;10:e0118212.
20. Pike R. Informed consent for medical photographs. Genet Med. 2000;2:353–5.
21. Enishi T, Yukata K, Takahashi M, Sato R, Sairyo K, Yasui N. Hypertrophic chondrocytes in the rabbit growth plate can proliferate and differentiate into osteogenic cells when capillary invasion is interposed by a membrane filter. PLoS One. 2014;9:e104638.
22. Crivellenti LZ, Cintra CA, Maia SR, Silva GEB, Borin-Crivellenti S, Cianciolo R, Adin CA, Tinucci-Costa M, Pennacchi CS, Santana AE. Glomerulotubular pathology in dogs with subclinical ehrlichiosis. PLoS One. 2021;16:e0260702.
23. Ng D. 2 types of digital images: tools to prepare stunning images. In: Bitesize Bio; 2023. https://bitesizebio.com/43785/an-introductionto-%20digital-images-in-publications/.
24. Vaidyanathan A. Images in scientific writing. J Indian Prosthodontic Soc. 2022;22:107.
25. Research Guides: All about images: Image File formats. https://guides.lib.umich.edu/c.php?g=282942&p=1885348
26. Kunz TC, Rühling M, Moldovan A, Paprotka K, Kozjak-Pavlovic V, Rudel T, Fraunholz M. The expandables: cracking the staphylococcal cell wall for expansion microscopy. Front Cell Infect Microbiol. 2021;11 https://doi.org/10.3389/fcimb.2021.644750.

Referencing

8

Dania Imtiyaz Khan, Islam Mohammad Shehata, Omar Viswanath, and Jamal Hasoon

8.1 Definition

Referencing is the practice of acknowledging "the sources of knowledge" used in any academic writing.

It is a standardized process that involves **citing** these sources in a consistent and structured manner within the text and providing a detailed **list of sources** at the end of the paper [1–3].

8.2 Brief Dive into History

Referencing, in its simplest form, began when scholars and thinkers chose to acknowledge the sources of their knowledge, usually through direct mentions or attributions within the text [4].

Citations are the currency of academic credibility.

D. I. Khan (✉)
Faculty of Medicine, Ain Shams University, Cairo, Egypt

I. M. Shehata
Department of Anesthesiology, Faculty of Medicine, Ain Shams University Cairo, Cairo, Egypt
e-mail: Islam.shehata@med.asu.edu.eg

O. Viswanath
Department of, Anesthesiology, Creighton University School of Medicine, Phoenix, AZ, USA

Mountain View Headache and Spine Institute, Phoenix, AZ, USA

J. Hasoon
Department of Anesthesiology, Critical Care, and Pain Medicine, University of Texas Health Science Center, Huston, TX, USA
e-mail: Jamal.J.Hasoon@uth.tmc.edu

© The Author(s), under exclusive license to Springer Nature Switzerland AG 2025
I. M. Shehata, O. Viswanath (eds.), *How to Successfully Publish a Manuscript*, https://doi.org/10.1007/978-3-031-92538-2_8

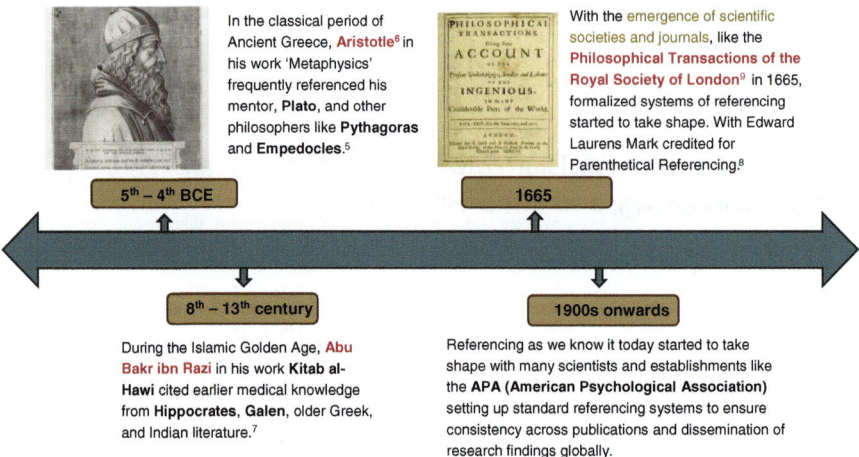

Fig. 8.1 Brief timeline of referencing

Fig. 8.2 Importance of referencing

These early references aimed to honor the contributions of predecessors and provide intellectual context but lacked a structured system; from that to modern-day referencing has been a long journey, and a few examples highlighting this evolution are mentioned in Fig. 8.1 [5–9].

8.3   Importance of Referencing (Fig. 8.2) [3, 10, 11]

8.4 Adding References

8.4.1 When Should You Add References? [3, 11]

Yes	No
Each time you use someone else's work or your own previously published work	Fundamental common knowledge like facts, dates, and information that can be checked in numerous publications and is not in dispute. (e.g.: normal blood pressure, etc.)
All material available on the internet, including illustrations	If you are summarizing ideas already introduced in the manuscript you dont need to add references again
If sharing personal or anecdotal sources you must be sure to reference whos experience you are sharing and back them up with substantial scholarly material	

8.4.2 How Many References Should You Include?

There is no universal standard for the number of references, but one should *consult the journal's author guidelines for specific limits* as there can be restrictions on the maximum or minimum number of references.

Refrain from

- Cluttering the references with unpublished data or partially relevant material.
- Including vague/ irrelevant self-citations solely to increase your citation counts (self-inflation).

8.4.3 Using the Right References

8.4.3.1 What Are the Right References?
- Recent (preferably in the last 5 years to reflect the latest findings).
- Having multiple citations.
- Published by reputable medical journals, organizations, or institutions.
- Undergone peer review, which ensures other experts have evaluated the research.

8.4.3.2 What Sources to Use? (Fig. 8.3) [3, 12, 13]
Peer-reviewed articles, Books, Magazines, and Newspapers are all reliable sources. On the other hand, *Wikipedia, blogs, Q&A sites (like Quora or Stack Exchange), and YouTube* are not acceptable sources to use directly as references.

- Wikipedia cites its references under each page which could lead you to better and more credible reading material.
- On Q&A forums, answers from credible staff members can be used as references.

Fig. 8.3 Sources for research papers

8.4.3.3 Navigating Through Different Sources

- If you are starting an academic paper, you can gather reliable references directly from the sources mentioned above; however, *other websites/ search engines* can make this process easier:
  1. **Semantics Scholar** is an AI-powered search engine that can help you navigate through research papers. It offers:
     - Better search filters.
     - Citation graph and influence metrics, which help you identify highly impactful papers and track the development of ideas across the specific literature.
     - Under the References tab of a paper, you can find direct links to its cited sources, providing valuable reading material sources [14].
  2. **Connected Papers** is a website that can be useful to find related papers. It features a unique tool that allows you to view a diagram of related papers and their time stamps. To use it, simply search for your paper and click the connected dots button. This is a quick way to find related articles and visualize their connections [15].
  3. **Google Scholar** is one of the simplest search engines available to access academic papers. Here is a short description of how to navigate through Google Scholar [16] (Figs. 8.4 and 8.5).
- You start by searching for the relevant topic. A few ways to *make the search process efficient are as follows:*
- Enter the relevant text in **"quotation marks"**, and Google performs an exact word-to-word search. For example, *"Genetic mutation in cystic fibrosis"*—this search input will return results that contain this exact phrase.

8 Referencing

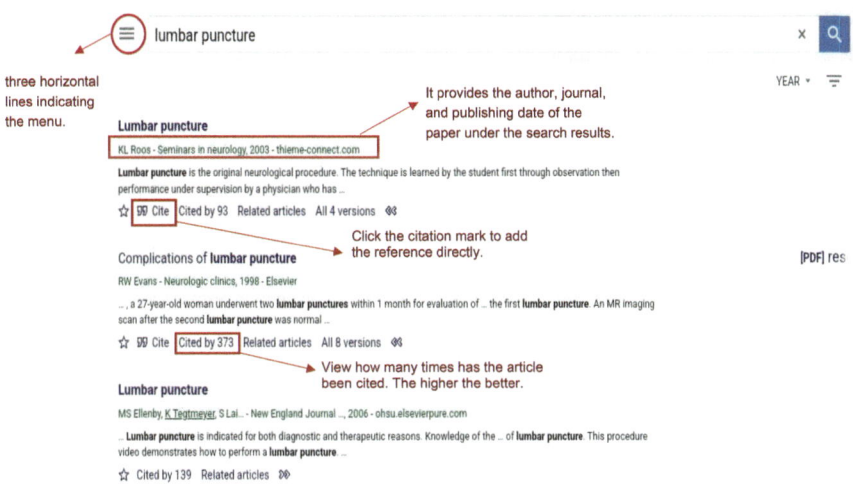

Fig. 8.4 Various features of Google Scholar

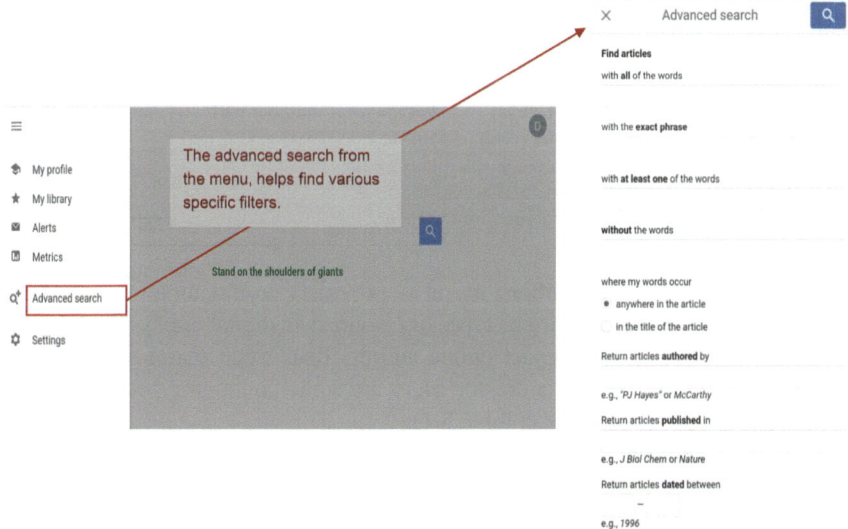

Fig. 8.5 Filters in advanced search

- Use **"OR"** and **"AND"** as logical operators. For example, *"cardiovascular disease" AND "hypertension"*—this search will return results that include both cardiovascular disease and hypertension. **"diabetes" OR "insulin resistance"**—The results will be related to either diabetes or insulin resistance.
- **Minus sign (−)** will exclude a term or search phrase. For example, *"heart disease" - "diabetes"*—this will exclude any results related to diabetes in your search for heart disease.

Table 8.1 Resource locators / identifiers

Type	Full form	Definition	Examples
URL	Uniform resource locator	A unique string of characters or an address, to identify or access a resource on the internet. Used extensively to share resources [17]	*http://www.randomexample.com/index.html*, which includes a protocol (HTTP), a domain name (www.randomexample.com), and a file name (index. Html) or the path to the page
DOI	Digital object identifier	A unique alphabetic and/or numeric string allocated to any digital file like a journal article, conference paper, report, or dataset, providing the exact link to its location on the internet [18]	*10.1000/182* or it can start with *https://doi.org* and then the respective numbering associated with the page
ISBN	International standard book number	A unique number intended to be used commercially, on the internet, and for library purposes to identify books and their specific editions. This can be used to add the exact book reference to your reference manager [19]	It can be **ISBN-10** having 10 digits and 4 sections, e.g., *0–061-96,436-7*, and a newer **ISBN-13** with 13 digits and 5 sections introduced after 2007, to have more available numbers, e.g., *978–0–061-96,436-7*
ISSN	International standard serial number	A unique eight-digit number to find a serial publication, such as a journal or magazine [20]	*1234–5678*, the last digit is a check digit and can sometimes be the letter X; e.g., *1050-124X*
PMID	PubMed identifier	A number assigned by the National Library of medicine to every paper listed on PubMed (search engine)., facilitating easy referencing and access to biomedical literature [21]	PMID: *12345678*, 1 to 8-digit number with no leading zero

- **Resource locators/ identifiers** are also extremely useful tools to accurately identify, access, and cite various types of resources in digital or physical formats. A resource locator is an exact online address that, when searched, takes you directly to the resource it codes for. Here are some common types of resource locators and their definitions (Table 8.1):
- **Secondary referencing** refers to the practice of citing sources mentioned in the text of another source, rather than consulting and citing the original source directly. It is generally advised to avoid this practice [3, 4].
- For example, if you find relevant information crediting another source in an article, blog, or review online, locate and cite the original source rather than relying on the website.
- Once you you have found your exact reference online, you can save it directly while viewing the webpage using a reference manager. Further details are provided in the sections below.

8.4.4 Citation

Citation is the process of adding a reference to your manuscript. It includes:

Numbering:
It is the most used type of citation.
Format: Text relevant to citation **(1)**—in the reference list, you will find the relevant source listed under number 1
E.g., Studies have shown significant reduction in complications with early administration **(1)**

Parenthetical Citations:
Format: **(Author, Year)**
E.g., The case progresses **(Dania, 2024)**

Multiple Authors:
Format: **(Author 1 & Author 2. Year)** or **(1st Author et al., Year)**
E.g., **(Delara & Dema. 2024)** or **(Imtiyaz et al., 2024)**

Quotations:
Direct excerpts from a source, enclosed in quotation marks, which use the source's exact wording
Format: **(Author, Year, p. Page Number)**
E.g., "The case progresses" **(Tabia, 2024, p. 37)**
[Square brackets] are used within the quotation marks to make edits. This can help add context or clarify parts of a quotation,
E.g., "They reported a significant increase in cases over the past decade." changed to "They reported a significant increase in [asthma] cases over the past decade."
Using**"[sic]"** in quoted citations is used to indicate that an error or unusual phrasing is part of the original source, not a transcription error
E.g., a misspelling of "diagnosis" as "diagnosis," adding "[sic]" after the misspelling shows that the error comes from the original text

- **In-Text Citation**: Include a citation to indicate where you have used the source. They can be written in multiple formats [22].
- **Reference List:** This section provides complete details for each source at the end of your document. It includes the author's name, the title of the work, the publication date, the publisher, and any other necessary information, all arranged according to the specified referencing style. The format will vary for books, journal articles, websites and other sources [23].
  - *Referencing styles* are standardized methods for citing sources in written work, dictating the format of citations and bibliographies, and ensuring consistency and clarity. As mentioned in Table 8.2, there are various styles, each with a different syntax [3, 11, 24].

Table 8.2 Different referencing styles

Style	In-text citation	Reference list entry	Examples
Vancouver style	Sequential number in (parentheses) or superscript [1]	**Book:** Author surname (s). Title of book. Edition (if it isn't first edition), place of publication: Publisher; Year.URL/DOI if available. **Journal Article:** Author surname (s).: Title of article. Abbreviated Title of Journal. Year Month Day; Volume (Issue). Page numbers. **Website:** Author surname (s). Title of webpage. Abbreviated title of journal, [Internet] published date. Volume (Issue). Page no. URL/ DOI	*Springer* *The New England Journal of Medicine*, *The Lancet*
AMA (American Medical Association) style	Sequential number in superscript	**Book:** Author surname (s). *Book title* edition. Publisher; year. URL/ DOI if available. **Journal Article:** Author surname (s). "Title of article." *Journal Name*. Volume (Issue). Year, pages (if online, DOI) **Website:** Author surname (s). "Title of webpage". *Title of the website* published date/ date of last update, URL	*Journal of the American Medical Association (JAMA)*
APA (American **Psychological Association) style** [23]	Parenthetical: (Author, Year) Narrative: Author (Year)	**Book:** Author surname, A.A. (Year). *Title of book*. Edition (if not first) Publisher. DOI if available **Journal Article:** Author surname, A.A. (Year). Title of article. *Title of Journal, volume number* (issue number), pages. If online https://doi.org/xx.xxx/yyyy **Website:** Author surname, A.A. (Date the site was published/ updated). *Title of webpage*. Site Name. URL/ DOI	*The Journal of Health Psychology*

(continued)

Table 8.2 (continued)

Style	In-text citation	Reference list entry	Examples
Harvard style	Parenthetical: (Author, Year), 'page number' when it is a direct quote. Narrative: Author (Year)	Book: Author surname, A.A. (Year of publication). *Title of Book*. Edition (if applicable). Place of publication: Publisher Journal Article: Author surname, A.A. (Year). 'Title of article'. *Journal Name*, volume number (issue number), pages. {If online https://doi.org/xx.xxx/yyyy} Website: Author surname, A.A., (Year). 'Title of webpage'/ Site Name. [version] available at URL (Accessed: Date).	Sometimes, *The BMJ (British Medical Journal)*

8.4.4.1 Steps to Create a Citation

1. Identify the Source: Determine what type of source you are citing (book, article, website, etc.).
2. Collect Information: Gather all necessary details (author, title, publication date, publisher, etc.).
3. Follow Citation Style: Use the appropriate format based on the style required (APA, MLA, etc.).
4. In-Text Citation: Insert the brief form of the reference within your text where you mention the source.
5. Reference List Entry: Add the full form of the reference in a reference list at the end of your document. Arranged in the order specified by the referencing style.

8.4.4.2 AI Tools to Create Citations

1. **Scispace:** https://scispace.com/
2. **CiteDrive**, syncs BibTeX files for LaTeX users and suggests accurate citations: CiteDrive
3. **RefWorks**, organizes references and integrates with databases like PubMed: RefWorks
4. **EasyBib**, automatically generates citations by scanning books, articles, and websites: EasyBib

8.4.4.3 Bibliography

It is a comprehensive list of all the sources referenced or consulted during the research and writing process of an academic paper or book. It includes the reference list and additional sources used for background reading.

- Depending on the *referencing style*, it can be written alphabetically based on the author's name or numerically.
- Other *specific details* like the spacing between entries or the use of abbreviations for journal names must be checked in the *journal's author guideline*.
- An *annotated bibliography* includes a summary or evaluation of each source [3, 4, 11].

8.4.4.4 Referencing Images and Figures

Images and figures can include a wide variety of visual content like photographs, diagrams, charts, graphs, and tables. When using such material in research, proper attribution is required, as most visual materials are protected by copyright [3, 11]. This necessitates the use of licensing. It is the legal permission granted by the copyright holder that defines how their work can be used, shared, or modified [25]. The most commonly used licenses are mentioned below:

- All rights reserved: The image or figure cannot be used without permission.
- Creative Commons licenses: These licenses allow reuse under certain conditions, such as attribution to the creator, noncommercial use, or nonmodification.
- Public domain: The image is free to use without any copyright restrictions [25].

For direct use, the image must be cited exactly as it is, while adaptations require a clear indication of modifications along with proper attribution to the original source.

The **referencing format** includes an in-text citation or a caption acknowledgment, following journal requirements and the chosen style, to specify whether the image is reprinted, reproduced, or adapted. The full reference must also be included in the reference list or bibliography as per the required style.

8.5 Organizing References

8.5.1 Reference Managers

A research paper can include over a hundred references and citing them by hand, in the correct format and style can be an extremely exhausting task. This is why we use programs called Reference Managers. These are specialized software designed to enable users to:

1. Efficiently COLLECT references and all related information from diverse sources by:
    - Directly importing them from online databases, library catalogs, academic journals, and websites using **browser extensions and plugins.**
    - Manually adding them using their ISBN, DOI, PMID, ADS Bib codes, or other resource locators.

2. To ORGANIZE references
   - In a structured library accessible for all your papers.
   - By categorizing into folders, subfolders, and tags.
   - For easy retrieval by **adding tags and annotations** to individual references.
   - Locating and **deleting duplicate references.**
3. ADD CITATIONS effortlessly by
   - **Integration with Word Processors** like Microsoft Word, Google Docs, LaTeX, etc.
   - A tab on the taskbar helps **directly insert in-text citations** into documents, in a wide range of citation styles (customizable).
   - Create a bibliography, which is automatically updated with each new reference added.
4. COLLABORATE
   - Sharing reference libraries along with annotations makes it easier for all authors to use a single library and cross reference.
5. SYNC reference libraries across multiple devices and access references and annotations from anywhere with internet connectivity.

You can download the *REFERENCE MANAGER APPLICATION* and its *BROWSER CONNECTOR* from its website.

- The main application is where all your references are stored and organized.
- The connector is the link between the browser and your main application. It shows up on the top right of your browser taskbar.

Examples of Reference Managers are Zotero, Mendeley, EndNote, RefWorks, BibDesk, JabRef, and many more.

- The software **Zotero** is an *open-source software* that is easy for beginners to learn. It integrates seamlessly with web browsers through its browser plugin Zotero Connector. Supporting all major platforms, Zotero offers great features for organizing, annotating, and sharing references [26].
- **EndNote** is a powerful tool with extensive database import capabilities and deep integration with Word Processors. It is especially popular among academics who require *access to institutional databases*. Its web importer is called is endnote click [27].
- **Mendeley reference manager** was originally part of the Mendeley desktop platform which is now obsolete. The available software consists of the main application, the web connector known as the Mendeley web importer, and Mendeley cite for integration with Word Processors. It features a ***user-friendly interface and great organization tools*** [28].

8.5.2 Comparison of Different Reference Managers (Table 8.3)

The software versions listed in Table 8.3 are current as of November 15, 2024.

8.5.3 Additional Pointers

- **Integration with Word Processors:**
  - Reference managers like Zotero, Mendeley, and EndNote integrate with Word Processors using plugins that *add bookmarks or dynamic fields that contain hidden metadata*, like author names, publication year, title, and a unique reference ID linked to the manager's library.
  - This *allows automatic updates to metadata* and style changes while maintaining proper formatting. This *allows automatic updates to metadata* and style changes while maintaining proper formatting.
  - In Microsoft Word, a field code system is used to store and interpret this data, Visible citations appear as formatted text (e.g., *Dania et al., 2020*) while underlying codes (e.g., {ADDIN ZOTERO_ITEM CSL_CITATION …}) store the metadata. You can *toggle the visibility of these fields by pressing Alt + F9 to check for errors in the citation style*.
- **Therefore:**
  - When sharing documents, it is important to *ensure collaborators have access to the same reference manager and library to retain dynamic links*. Without access, citation fields may convert to plain text, breaking the link to the library. To avoid such issues, *use shared groups and compatible platforms*, or share finalized documents *by converting citations to plain text* to avoid any errors in formatting.
  - Always **keep a master version of the document with active fields** for future updates or use a *shared Google Document* (especially compatible with Zotero).
  - For troubleshooting, check for missing plugins, mismatched citation styles, or whether Word's live collaboration mode is enabled, as this may disable field codes.
  - Unfortunately, once citations in a document are converted to plain text, the dynamic link to the reference manager's library is lost, and there's no straightforward way to "reactivate" them.
- Exported citations can lose formatting or data during transfer between managers, so make sure to **use proper formats like BibTeX or RIS and verify metadata post-import.**
- Annotations created in the built-in PDF reader are stored in the reference manager's database, so they won't be visible in external PDF readers unless you *export a PDF with embedded annotations*.
- To use the browser connector, you typically need to have the application running in the background.
- If you are unable to save metadata from a particular website, most likely the page is not supported by the browser plugin.

Table 8.3 Comparison of different Reference Managers

	Zotero 7.0.9 [29, 30]	Endnote 21.4 [31]	Mendeley Reference Manager V2.125.2 [32]
Links to download	https://www.zotero.org/download/	https://endnote.com/downloads/	https://www.mendeley.com/download-reference-manager/
Availability	Free (paid for extra storage)	Endnote Online/ Endnote Basic is free but lacks many features. The complete desktop version is paid. (Many universities provide free access)	Free, requires you to have an Elsevier account, which is also completely free. Paid plan for more storage and access to larger groups
Platform (OS)	Windows 7 onward, Mac OS 10.12 or later, Linux	Windows 10,11, Mac OS 10.14 onward	Windows 7 onward, Mac OS 10.10, Linux
Browser integration	**ZOTERO CONNECTOR** Chrome, Edge, Firefox, and Safari (there are a few limitations on Safari)	**Endnote click** Chrome, Firefox, Edge, and Opera. (Weaker integration)	**Mendeley web importer** Chrome, Firefox, Safari
Word processor integration	Microsoft Word, LibreOffice, *Google Docs*. Adds reference using the Zotero tab	Microsoft Word, *Apple Pages*, Apache OpenOffice, LibreOffice. Adds reference using the endnote tab	Microsoft Word 2016 and above, LibreOffice. Adds references using *Mendeley Cite* (Makes the process easier)
Cloud storage	Free 300 MB, paid plans for more	2 GB free, paid plans for more	2 GB free, paid plans for more
Adding references	Manually, using identifiers, uploading PDFs, and using Zotero connector (**which integrates very smoothly**) Add several types of references including books, journals, interviews, and podcasts. (offering the most diverse options available) Can import a library directly from Mendeley reference manager (online import) If PDFs are open access, they will be downloaded	Manually, using identifiers, uploading PDFs, using Endnote click There is an **ONLINE SEARCH FEATURE** available in the application, which *provides access to multiple databases and university library catalogs* If PDFs are open access they will be downloaded	Manually, using identifiers, uploading PDFs, and using Mendeley web importer You can search for sources online at mendeley.com and directly add the references Includes a feature called **WATCHED FOLDER**, *which allows the user to select a folder on their desktop and Mendeley automatically adds all references saved in that folder to the library and updates the library automatically as the user adds more references* If PDFs are open access they will be downloaded

(continued)

Table 8.3 (continued)

	Zotero 7.0.9 [29, 30]	Endnote 21.4 [31]	Mendeley Reference Manager V2.125.2 [32]
Organizing reference	Enables the creation of folders and subfolders in your library Includes dedicated sections for my publications, trash, unfilled references, and duplicate references Features an advanced search option that allows searching with multiple filters	Enables the creation of sub-sections in the library under my groups Includes dedicated sections for my publications, trash, unfilled references, and recently added references The **SMART GROUPS** feature enables users to *define specific search criteria for a group*. Any references that meet these criteria are automatically added to the group. These groups remain updated as the user adds more references to the library	Allows creating collections in your library Has additional *sections, including favorites, and recently read*; along with recently added, trash, unsorted, and my publications Has a *filter option* to view the library through certain filters
Collaboration/sharing	Create, share, and collectively edit, and annotate references in a group. They can be: Public-Open, Public-Closed, or Private Membership Allows transfer of group ownership **UNLIMITED MEMBERS** are allowed in shared libraries	Create, share, and collectively edit, and annotate references in shared groups Up to 1000 members	Create, share, collectively edit, and annotate references in groups Allows transfer of group ownership 25 members per group for free accounts, 50 members per group for accounts with paid storage, and 100 members per group for institutions
PDF annotation (notes, comments, etc.)	Yes, built-in along with annotations for EPUB and webpage snapshots Allows highlighting, underlining, and creation of notes Can add tags	Yes, built in With sticky notes, highlight, underline, and strikethrough Can add tags	Yes, built in Offers a **NOTEBOOK FEATURE** that enables *easy reading and annotating of PDFs, providing more diverse functionality*. These notes are not required to be linked to a specific reference Can add tags

8 Referencing

Citation styles	Thousands of styles available, customizable	Thousands of styles available, customizable	Thousands of styles available, customizable
Duplicate detection	Yes	Yes	Yes
Full-text search	Yes, within PDFs, tags, and notes	Yes, within PDFs, tags, and notes	Yes, within PDFs, tags, and notes
Metadata retrieval	Automatic retrieval and updating	Automatic retrieval and updating	Automatic retrieval and updating
Export formats	Zotero RDF, BibTeX, RIS, CSL JSON, RefWorks tagged, EndNote XML, and others	BibTeX, RIS, EndNote XML, and others	BibTeX, RIS, EndNote XML, and others
Mobile app	iOS, Android (third-party apps)	iOS only (EndNote for iPad)	iOS, Android
User support	Community forums benefit from extensive documentation due to their active user bases and the flexibility of open-source software	Community forums, official support, dedicated customer service, and tailored institutional training	Community forums: Mendeley blog, official support, and collaborative solutions with group-based assistance, (individual help may be slower
Extra features	As a free, open-source software, the user community continuously improves it. Integrates well with various tools, such as research rabbit and AI. **Zotero** is ideal for students and researchers needing free, versatile management across platforms	It automatically creates a bibliography, unlike other software that requires manual addition. Powerful search capabilities and the ability to create complex queries to filter and organize references. **EndNote** is tailored for detailed and high-volume academic projects but comes with a steep cost	**Mendeley** shines in PDF-centric workflows and group collaborations

8.6 Common Mistakes to Avoid

1. **Collecting your references**
   (a) As you gather information, *save and organize* your sources with brief context notes. This ensures you do not lose track of your original sources.
   (b) For *shared projects*, keep a separate document for each contributor's manuscripts with their references intact to maintain consistency and facilitate cross-checking.
2. **Missing citations**
   (a) When managing citations manually, double-check that every in-text citation has a matching reference in your paper.
   (b) Use a *color-coding system* while drafting to organize better:
      (i) Green for references that are cited and added to the list.
      (ii) Blue for references cited but unlinked.
      (iii) Red for incomplete citations.
   (c) Use placeholders, like "(source needed)," to remind yourself to add the missing citation.
3. **Improper formatting** can include any problem with the way the citation or reference is written; it can be:
   (a) Missing data, like publication year/ place/ page numbers/volume.
   (b) Omitting the web address when required.
   (c) Incorrect punctuation.

   The easiest way to avoid minor citation errors is by using *reference managers, AI tools, or direct citation features from platforms like Google Scholar* (as mentioned above). If you are citing manually or need to cross-check, refer to the **PUBLISHED GUIDELINES** to ensure accuracy:

   - **Harvard referencing** varies by publication but *Cite Them Right: The Essential Referencing Guide (12th edition) by Richard Pears & Graham Shields* [3] is widely regarded as the gold standard.
   - **Vancouver style** follows *Citing Medicine: second edition, The NLM Style Guide for Authors, Editors, and Publishers by Karen Patrias and Dan Wendling* [33], endorsed by the International Committee of Medical Journal Editors (ICMJE) [34].
   - **APA style** has *the Publication Manual of the American Psychological Association (7th edition, 2020)* [35] or its official guide.
4. **Uncommon sources**

   For sources like Conference proceedings, Newspaper articles, Audio and Visual Media, Citing Material on the Internet (Online), etc., refer to the **specific referencing guidelines** (mentioned above) to find the correct citation syntax.

   For PDFs, find the associated book, journal, or author (personal or organization) and reference it in its original form or as an online webpage (use the syntax for books/articles, etc. or a webpage, respectively).

References

1. References [Internet]. https://apastyle.apa.org. [cited 2024 Nov 11]. Available from: https://apastyle.apa.org/style-grammar-guidelines/references
2. Library TU of Q. 3. Referencing. 2023 Jun 29 [cited 2024 Nov 23]; Available from: https://uq.pressbooks.pub/digital-essentials-write-cite-submit/chapter/referencing/
3. Pears R, Shields GJ. Cite them right: the essential referencing guide. 12th ed. New York: Bloomsbury Academic; 2022. p. 279. (Bloomsbury study skills)
4. Gastel B, Day RA. How to Write and Publish a Scientific Paper. 9th ed. Erscheinungsort nicht ermittelbar: Greenwood; 2022. p. 1.
5. Aristotle. Metaphysics. NU Visions Publications; 2009. 215 p.
6. Anonymous, Italian, mid-16th century, After Enea Vico (Italian, Parma 1523–1567 Ferrara). Aristotle, from "Speculum Romanae Magnificentiae" [Internet]. 1553 [cited 2025 Mar 6]. Available from: https://www.metmuseum.org/art/collection/search/370768
7. Ahmed JO, Kakamad KK, Najmadden ZB, Saeed SI. Abu Bakr Muhammad Ibn Zakariya Al-Razi (Rhazes) (865–925): the founder of the first psychiatric ward. Cureus. 16(7):e64601.
8. McElligott M. Lib Guides: Reference & Citation: History of Harvard Style [Internet]. [cited 2025 Feb 23]. Available from: https://ncad.libguides.com/c.php?g=688721&p=4928704
9. Royal Society (Great Britain), Britain) RS (Great, Hutton C, Maty PH, Pearson R, Shaw G, et al. Philosophical transactions of the Royal Society of London [Internet]. Vol. v.24=no.289–304 (1704–1705). London: Royal Society of London; 1704. 762 p. Available from: https://www.biodiversitylibrary.org/item/184346
10. Santini A. The Importance of Referencing. J Crit Care Med. 2018;4(1):3–4.
11. Neville C. The complete guide to referencing and avoiding plagiarism. 3rd ed. London: Open University Press; 2016. p. 229. (Open UP study skills)
12. Wecker E. Welch medical library guides: expert searching: which databases to use [Internet]. [cited 2024 Nov 23]. Available from: https://browse.welch.jhmi.edu/searching/databases-by-subject
13. 25 Best Health & Medical Information Sites [Internet]. RefSeek. [cited 2024 Nov 23]. Available from: https://www.refseek.com/directory/health_medical.html
14. Semantic Scholar | AI-Powered Research Tool [Internet]. [cited 2024 Nov 11]. Available from: https://www.semanticscholar.org/
15. Connected Papers | Find and explore academic papers [Internet]. [cited 2024 Nov 23]. Available from: https://www.connectedpapers.com/search?q=How%20to%20Search,%20Write,%20Prepare%20and%20Publish%20the%20Scientific%20Papers%20in%20the%20Biomedical%20Journals&p=1.
16. About Google Scholar [Internet]. [cited 2024 Nov 11]. Available from: https://scholar.google.com/intl/en/scholar/about.html
17. What is a URL? – Learn web development | MDN [Internet]. 2024 [cited 2024 Nov 11]. Available from: https://developer.mozilla.org/en-US/docs/Learn/Common_questions/Web_mechanics/What_is_a_URL
18. What is a digital object identifier, or DOI? (6th edition) [Internet]. https://apastyle.apa.org. [cited 2024 Nov 23]. Available from: https://apastyle.apa.org/learn/faqs/what-is-doi
19. What is an ISBN? | International ISBN Agency [Internet]. [cited 2024 Nov 11]. Available from: https://www.isbn-international.org/content/what-isbn/10
20. What is an ISSN? | ISSN [Internet]. [cited 2024 Nov 11]. Available from: https://www.issn.org/understanding-the-issn/what-is-an-issn/
21. PMID vs PMCID: What's the Difference? – NIH Extramural Nexus [Internet]. 2015 [cited 2024 Nov 11]. Available from: https://nexus.od.nih.gov/all/2015/08/31/pmid-vs-pmcid-whats-the-difference/
22. In-text citations [Internet]. https://apastyle.apa.org. [cited 2024 Nov 23]. Available from: https://apastyle.apa.org/style-grammar-guidelines/citations

23. Basic principles of reference list entries [Internet]. https://apastyle.apa.org. [cited 2024 Nov 11]. Available from: https://apastyle.apa.org/style-grammar-guidelines/references/basic-principles
24. Masic I. How to search, write, prepare and publish the scientific papers. Biomedical J Acta Inform Medica. 2011;19(2):68–79.
25. About CC Licenses [Internet]. Creative Commons. [cited 2025 Feb 24]. Available from: https://creativecommons.org/share-your-work/cclicenses/
26. Zotero | Your personal research assistant [Internet]. [cited 2024 Nov 23]. Available from: https://www.zotero.org/
27. Product details – EndNote [Internet]. [cited 2024 Nov 23]. Available from: https://endnote.com/product-details/
28. Mendeley Reference Manager | Mendeley [Internet]. [cited 2024 Nov 23]. Available from: https://www.mendeley.com/reference-management/reference-manager
29. changelog [Zotero Documentation] [Internet]. [cited 2024 Nov 20]. Available from: https://www.zotero.org/support/changelog
30. system_requirements [Zotero Documentation] [Internet]. [cited 2024 Nov 11]. Available from: https://www.zotero.org/support/system_requirements
31. Compatibility [Internet]. EndNote. [cited 2024 Nov 11]. Available from: https://endnote.com/product-details/compatibility/
32. Guides | Mendeley [Internet]. [cited 2024 Nov 18]. Available from: https://www.mendeley.com/guides
33. Patrias K, Wendling D. Citing medicine. 2nd ed. National Library of Medicine (US); 2007.
34. ICMJE | Recommendations for the Conduct, Reporting, Editing, and Publication of Scholarly Work in Medical Journals [Internet]. 2023 [cited 2025 Jan 25]. Available from: https://www.icmje.org/recommendations/
35. American psychological association, editor. Publication manual of the American psychological association: the official guide to APA style. 7th ed. Washington (D.C.): American psychological association; 2020.

How to Choose the Right Journal

9

Eman Hamdy Oweiss, Islam Mohammad Shehata, Omar Viswanath, and Rory Murphy

> *A well-chosen journal can amplify your voice; a poor choice can silence it.*
>
> —Anonymous

9.1 Why to Exercise Caution Upon Selecting the Journal

Preparing a manuscript is no simple mission. It is one that requires investing a great deal of time and effort. However, choosing the right journal is the one crucial decision that shall determine whether all your hard work will eventually pay off, as it is the vessel for your hard work. The journal that you choose will not only affect your paper's impact, but it has the potential to influence your own career as well.

- Impact on Your Paper
    1. Credibility

    The credibility of your research can be heavily judged by the prestigiousness and reputation of the journal where it is published. For instance, publishing in a

---

E. H. Oweiss (✉)
Faculty of Medicine, Ain Shams University, Cairo, Egypt
e-mail: 170950@med.asu.edu.eg

I. M. Shehata
Department of Anesthesiology, Faculty of Medicine, Ain Shams University Cairo, Cairo, Egypt
e-mail: Islam.shehata@med.asu.edu.eg

O. Viswanath
Department of Anesthesiology, Creighton University School of Medicine, Phoenix, AZ, USA

Mountain View Headache and Spine Institute, Phoenix, AZ, USA

R. Murphy
Department of Neurosurgery, St. Joseph's Hospital and Medical Center, Barrow Neurological Institute, Phoenix, AZ, USA

© The Author(s), under exclusive license to Springer Nature Switzerland AG 2025
I. M. Shehata, O. Viswanath (eds.), *How to Successfully Publish a Manuscript*, https://doi.org/10.1007/978-3-031-92538-2_9

"predatory journal" can deem your manuscript invalid regardless of the quality of your research.

2. Visibility

Visibility of your manuscript is highly dependent on the journal where it is published.
- Open-access (OA) journals have higher visibility than subscription journals. [1]
- Journals indexed in databases also have higher visibility than other journals [2]. Since visibility directly impacts the number of citations a paper receives, higher visibility will translate into a higher **H-index***.

H-Index

is a measure of an author's productivity as well as impact. It is calculated by arranging an author's publications from highest to lowest according to the number of citations; then, taking the number of publications that have been cited at least the same number of times. [3] The following table shows an example of how H-index is calculated for an author whose H-index is **7**. In that example, the author has nine publications. The seventh publication has been cited seven times. Since the eighth one has been cited less than eight times, H-index is estimated to be only 7.

Paper	Number of Citations
1	92
2	36
3	20
4	12
5	10
6	9
7	7
8	4
9	4

I10-Index

is another tool created by Google Scholar in 2011 to measure authors' productivity in a simpler manner. It is calculated by counting the number of publications that have received ten or more citations [4].

- *Impact on Your Career*
    1. Academic Advancement

   The number and quality of papers you publish give you leverage when it comes to academic promotions. It will also make you more competitive for higher-paying positions [5], securing grants, scholarships, fellowships, etc. [6] Moreover, being a well-known author gives you more visibility which increases your chances of receiving invitations to become an editorial board member or to publish in more journals.

    2. Connections

   Publishing in reputable journals helps you build your own reputation which increases your contact list of other peers [5].

9.2 When to Choose the Journal

"Which came first: the chicken or the egg?"—a dilemma that has perplexed humanity for centuries. But here is another dilemma for you: which comes first—writing a manuscript or choosing a journal? In this section, we will lay out different scenarios and tell you the pros and cons of each (Table 9.1).

Our Opinion

As a result of what we discussed earlier, we recommend that it may be much more appropriate to choose the journal after preparing a draft containing the basic structure of your paper. For that, you can use **IMRaD structure*** and write down the important points in each section. This way, you can have a primary outline that is malleable and easily manipulated to fit whichever journal you choose.

> **Journal Guidelines*:**
> Each journal has certain guidelines that could be found under the title "guidelines for authors" or "submission guidelines." These guidelines include criteria regarding the formatting of your paper. Most common examples of these criteria include the following:
> - Font specifications
> - Word count limit
> - Referencing style (Vancouver, Harvard, etc.)
> - Number of tables
> - Instructions regarding artwork (ex., number of figures)
> - Instructions for the structure used to lay out the body of the paper, i.e., variations of the IMRaD structure in case of study design [7].
>
> N.B. IMRaD structure is a structure used to lay out the body of a research paper after the title and abstract (Introduction, Materials and Methods, Results, and Discussion) [8].

Table 9.1 A comparison between writing the manuscript before and after choosing the journal

	Writing the manuscript first	Choosing the journal first
Pros	Liberal writing without being constricted to specific guidelines	Tailoring your manuscript according to the journal's target audience Writing readily according to the journal guidelines, saving lots of time
Cons	Requires extra time and effort as you will have to tailor your manuscript once again on the back end according *to* **journal guidelines*** [7]	Restricts you to journal guidelines while writing Necessitates that you re-tailor your manuscript from the start in case of rejection

9.3 How to Pick the Most Suitable Journal

Choosing a suitable journal is not a haphazard process. To guarantee maximum advantage for yourself, you have to follow a well-planned systematic approach (Fig. 9.1). In this section, we will break down this process into simple steps that can guide you along the way.

9.3.1 Phase of Listing Potential Journals

To make a list of all potential journals, YOU need to answer two main questions.

9.3.1.1 What Are Potential Journals?
By "potential journals," we primarily mean the journals whose scope conforms with your manuscript's content. One of the most common reasons for a manuscript to **get rejected** is if its topic does not fall under the scope of the journal. [9, 10] To avoid this, you must do the following:
- **Read the journal's "aims and scope" section.** This will give you an idea about the subjects covered by the journal as well as the type of articles accepted by it; some journals, for example, do not accept clinicals studies.
- **Investigate the journal's target audience.** Research suggests that you are likely to receive more citations if you choose a journal whose target audience matches your article's rather than a journal with a high impact factor [1]. Moreover, knowing the journal's audience allows you to decide on the language of your manuscript and whether you will need translation support.
  A journal's target audience can be defined in terms of specialization and geographic scope [7, 11].
- *Specialization*
    - General audience/ multidisciplinary
    - Specialized, single discipline

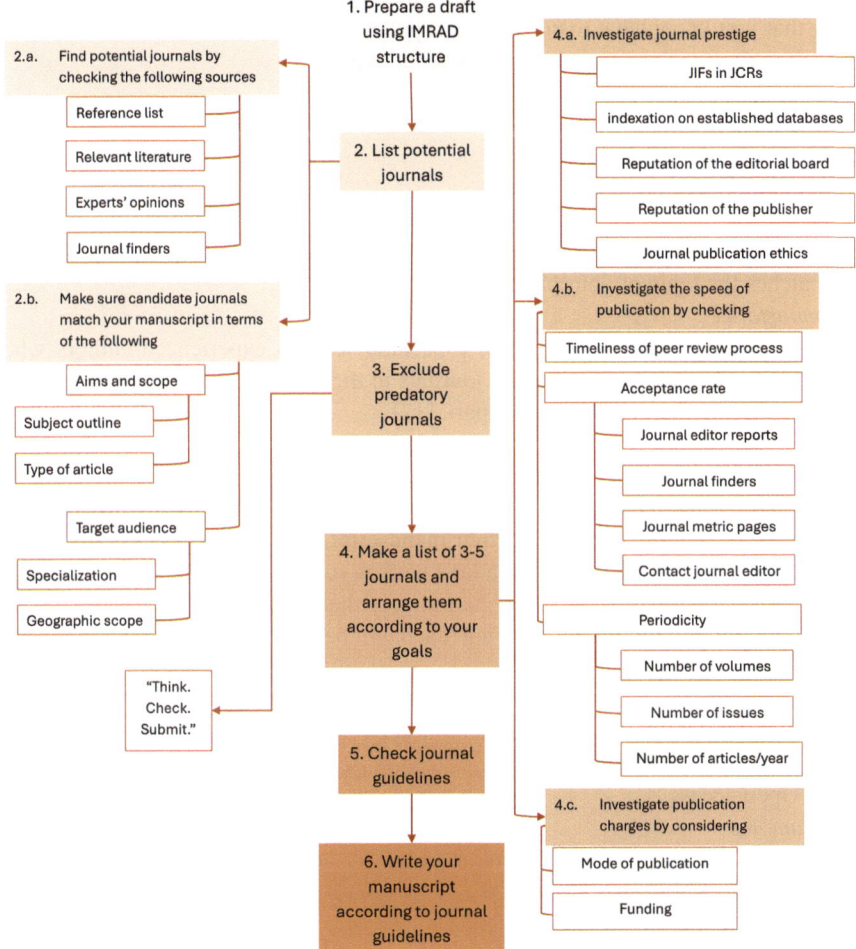

Fig. 9.1 A stepwise approach to picking the most suitable journal. *JCRs* Journal Citation Reports, *JIF* Journal Impact Factor

- *Geographical Scope*
  - International
  - National
  - Regional

An article on global health issues such as hypertension or diabetes is a good fit for an international journal, unlike an article discussing an endemic disease for which a national/ regional journal might be more suitable.

- **Read similar articles that were published recently by the journal**
- This will give you an idea about the topics in which the journal is currently interested as well as the type of articles it usually publishes.
- It will also let you know early on if the journal has published articles on similar topics to yours, which might lower your chances of acceptance. [1]

9.3.1.2 How Can You Find Potential Journals?

Now that you know what to consider in a potential journal, we need to further discuss the sources where you can find these journals:
- **Relevant literature**
- **Experts' opinions**
- **Your reference list**
- **Journal finders**
- *Definition*
  Journal finders are online services that generate a list of journals whose scope matches the topic of your paper.
- *Importance*
  They are highly convenient tools especially for less experienced authors who might not be familiar with many journals in their field yet. They offer their services free of charge for the most part. [12]
- *How They Work*
  They only require you to insert title and abstract +/− keywords+/− field of research. Then, it compares them with papers published by different journals to find the ones most suitable for your manuscript. These tools will often provide optional functions that you can adjust so that you find journals that match your priorities and goals as an author, not merely the subject of your paper [12, 13]. These options include the following (each will be explained later in this chapter):
  – Impact factor
  – CiteScore
  – Acceptance rate
  – Publication time
  – Open access
- *Examples*
  There are many journal selectors available for authors. The most popular ones include the following (Table 9.2).

9.3.2 Phase of Exclusion

Here, you must know that there are journals that should never be considered in the first place, namely predatory journals.

9.3.2.1 What Are Predatory Journals?

Predatory journals are journals you must weed out before starting to form your final list. The scientific community has no clear definition of what a predatory journal is [15]. Generally speaking, these are journals that require authors to pay APCs (Article Processing Charges) but do not offer the services that any reputable journal is expected to offer, namely, peer-review process. As a result, they usually publish plagiarized content or scientifically inaccurate manuscripts. Therefore, the quality and validity of any paper published in such journals will be compromised. [16]

Table 9.2 Examples of the most popular journal finding tools

Journal finder	Link	Additional notes
Elsevier journal finder	https://journalfinder.elsevier.com/	Elsevier and springer only suggest journals from their own library [12]
Springer journal suggester	https://journalsuggester.springer.com/	
Ednaz journal selector	https://en-author-services.edanzgroup.com/journal-selector	Edanz has the advantage of searching a vast array of journal publishers including Elsevier, springer, John Wiley & Sons, Taylor & Francis, and nature. It also offers a paid service where experts recommend four journals best suiting your manuscript [12, 13]
Wiley	https://journalfinder.wiley.com/	Wiley also offers a paid service similar to Edanz's [14]
The journal/author name estimator (Jane)	https://jane.biosemantics.org/	Jane mainly searches Medline database, which is published by NLM (National Library of medicine) [13]
Enago open access journal finder	https://www.enago.com/researcher-hub/journal-finder.htm	Enago searches DOAJ for indexed OA journals. It can also provide authors with a list of 3–5 journals best matching their manuscript. Moreover, it has the advantage of providing authors with feedback to improve their manuscript before submission [12]

9.3.2.2 Who Do They Target?

Predatory journals target mainly two groups of authors:

- Authors with little expertise in publication.
 These authors are particularly tricked into submitting to these journals while they still do not know how to properly check for the legitimacy of a journal.
- Authors with low-quality manuscripts who need publications for academic advancement. [17]

9.3.2.3 Why Should You Avoid Predatory Journals?

Given the fact that the articles published in predatory journals can mostly be scientifically inaccurate, their presence is considered a threat to the scientific community as a whole. On another level, these journals raise many concerns among authors. Once you have submitted your paper to a predatory journal, sometimes there can be no turning back even if you have not paid the APCs yet:

- You cannot submit your paper to another journal unless you withdraw it from the first one.
- Withdrawal can be a lot of trouble because you may never get a response from the predatory journal.
- If you have paid the APCs and, then, decide to withdraw your paper, there is a high likelihood that it will still be published and, thus, cannot be submitted to another journal. [18]

9.3.2.4 How Do You Avoid Them?
In order to identify predatory journals, you have to know the following:
A. *How predatory journals lure young researchers into submitting their papers.*
   Predatory journals usually send e-mails to young authors inviting them to publish at their journal, even after a single publication. They might guarantee acceptance of their paper and might also negotiate APCs, which is never the case with legitimate journals. [11, 16] As a new author, you have to realize that receiving an invitation to publish at a prestigious journal is very unlikely [19]. Therefore, it is safer to ignore these e-mails or, at least, check their eligibility—especially since journals might imitate the names of other reputable journals or use fake editor and/or reviewer names. [18]
B. *The characteristics of predatory journals.*
   - To check if a journal has any of the characteristics of a predatory journal, you can use the "Think, Check, Submit" checklist (https://thinkchecksubmit.org/).

   The checklist contains a group of questions that authors should answer regarding the journal they plan to submit their paper to. Major points covered by these questions include the following:
   - Journal reputation
   - Publisher contact information
   - Transparency regarding the peer-review process
   - Indexation
   - Transparency regarding APCs
   - Guidelines for authors
   - Recognized industry initiative membership (DOAJ, COPE, OASPA)
     - Another way to check if a journal is predatory is to go through blacklists and whitelists of journals.
   - Blacklists (lists of predatory journals).
     - Baell's List [21]
       Beall's list was the most popular list of OA predatory journals. [20] The original list made by Jeffrey Beall was taken down in 2017 due to legal issues. However, the old list is currently being kept and updated by an anonymous researcher.
     - Cabell's Blacklist
       Cabell's blacklist similarly enlists predatory journals. However, it includes both subscription and OA journals. It requires subscription charges.
   - Whitelists (lists of legitimate journals) [22]
     - DOAJ
       Directory of Open Access Journals provides a list of legitimate OA journals.
     - Cabell's Whitelist
       This list also requires subscription charges.

9.3.3 Phase of Shortlisting

After exclusion of certain journals, you will have to prioritize others in order to reach a final, refined list of —three to five journals. These journals should be arranged from top to bottom according to your publication objectives. As an author, your goals will most likely be one of the following:

- Publishing in a prestigious journal
- Rapid publication
- Low publication charges [2, 7]

In this section, we will discuss each of these goals in terms of when to prioritize them and what factors to look for in a journal to make sure it serves these goals (Table 9.3).

9.3.3.1 Journal Prestige

When publishing a manuscript, it is important to realize that its value is usually judged first by the journal where it is published rather than its actual scientific value [2]. That is why your career advancement could highly depend on journal reputation; for instance, it increases the competitiveness of your total body of work. Moreover, it increases your chances of securing future funding opportunities and receiving invitations from other reputable journals. In order to find these top-ranking journals, you need to look for the following factors:

I. Citation-Based Metrics
   As the name implies, "citation-based metrics" is a measurement of journal impact that is calculated based on the number of citations that its papers get. These include the following:
   - Journal Impact Factor (JIF)

   JIF is the most popular indicator of the journal prestige. [23] It has been used since 1975. [24] Until this day, JIFs are published annually in Journal Citation

Table 9.3 Possible publication goals and how to investigate them

Goals	Motives	Factors
Publishing in prestigious journals	Academic career advancement More publication opportunities	Citation-based metrics; most commonly, impact factor Indexing by established databases Peer-review process and reputation of the editorial board Reputation of the publisher Journal publication ethics
Rapid publication	Career advancement Significance of results	Timeliness of the peer-review process Acceptance rate Periodicity
Low publication charges		Type of journal/mode of publication Funding

Reports (JCR) by Clarivate Analytics, formerly Thompson Reuters. [25] JIF is calculated by dividing the number of citations received by a journal in the past 2 years by the number of citable items published in those years. [26] The term "citable items" is considered a little vague, which is a point of criticism of JIF; however, it typically includes the original reports and reviews but excludes editorials, letters to editors, new items, tributes and obituaries, correction notes, and meeting abstracts [27].

$$JIF = \frac{\text{number of citations in the past 2 years}}{\text{number of citable items published in the past 2 years}}$$

This gives an idea of the average number of citations received by any article during the preceding 2 years; thus, it roughly estimates the impact and distribution of the paper [26].

In spite of its popularity, JIF has to be used with caution as it faces a lot of criticism and controversy [6]. This criticism is caused by the fact that JIF can be affected by factors that have nothing to do with the quality of the journal. These factors include the following among others:

- The type of articles published by a given journal—for instance, review articles, especially systematic reviews, are typically cited a lot more frequently than case reports [25].
- The timing of publication for citable items: The earlier citable items are published in the span of these 2 years, the more time they have to receive more citations [25].
- The definition of "citable items" in each journal (the denominator) and inflation of the numerator by "free citations," i.e., citation of items excluded from the denominator [27].
- Journal specialty: This makes JIF of limited use in particular fields and in assessing multidisciplinary journals. It also means that cross-field comparison will not make sense [27]. However, this can be overcome by comparing JIF quartiles where journals in the first quartile are the 25% top journals in the field and vice versa [2].
- JCR coverage bias in favor of journals published in the English language: This somehow gives an advantage to journals from certain nations over others for reasons other than quality [27].

  In conclusion, it is important to consider IF when choosing a journal, but keep in mind that it is not a completely accurate indicator of journal quality.
- CiteScore
- SCImago Journal Rank (SJR)
- Source Normalized Impact per Paper (SNIP) [1]

II. Indexing in Established Databases
   - What does it mean for a journal to be indexed in a database?

An index—as defined by Merriam Webster—is "a list (as of bibliographical information or citations to a body of literature) arranged usually in alphabetical order of some specified datum (such as author, subject, or keyword)." Likewise,

a journal being indexed in a database means that this journal has met the criteria of this database and is, therefore, indexed.
- Why is it important to make sure that a journal is indexed?
  - Quality: Just like a paper gets reviewed by journals to make sure it has valid scientific data, a journal has to meet certain criteria set by each database to make sure it has a sound review process [28]. Therefore, journal indexation is a very important indicator of quality. And, as mentioned before, checking if a journal is indexed is a useful means of excluding predatory journals.
  - Visibility: Being indexed in established, well-known databases means that your paper will likely be viewed more frequently [1].
- What are the most reliable databases?
  - MEDLINE (NLM) and PMC

(Note that there are articles in PubMed that are published in journals not indexed for MEDLINE; therefore, you have to check the NLM catalogue and make sure the journal is labeled "currently indexed for MEDLINE.")
  - EMBASE
  - SCOPUS
  - Web of Science (Clarivate)
  - DOAJ
- How to check if a journal is indexed in a certain database?
  - Go to the database website.
  - Search by journal name or ISSN (International Standard Serial Number).

III. Peer-Review Process and Reputation of the Editorial Board

Indicators that a journal has a sound peer-review process include the following [29]:
- Transparency regarding the number of reviewers for each paper and whether the journal recruits external reviewers.
- Availability of editors' contact information.
- Reputation and expertise of the editorial board.
- Transparency regarding timeline from submission to acceptance/rejection without promises of assured acceptance.

IV. Reputation of the Publisher

To name a few, reputable journals include the following among others [19]:
- Elsevier
- Springer
- SAGE
- Lippincott
- Taylor & Francis
- PLOS One
- NEJM
- BMJ
- John Wiley & Sons
- Wolter Kluwer
- Informa
- MedKnow

V. Journal Publication Ethics

This can be verified by checking whether the journal has membership in different ethics organizations [13]:
- COPE
- Association of STM Publishers
- ICMJE
- OASPA

More details will be further elaborated in the next chapters.

9.3.3.2 Timeliness of Publication

Although publishing in a prestigious journal is a key consideration to nearly all authors, there will be times when you will have to balance this desire with the need for rapid publication:

- Some promotions will dictate that you publish a manuscript before they can be granted to you. This means that you must consider journals that can offer rapid publication so that you can meet the deadline set for that promotion [30].
- Your results can be more significant within a particular timeframe, which might motivate you to favor rapid publication to a certain extent [30]. An example of this is the times of COVID-19 pandemic when the rate of emergence of new data was extremely high and the need for new scientific input was at its peak.

If those cases apply to you, then you should consider the following factors:

I. Timeliness of the Peer-Review Process

The peer-review process is mainly composed of the following steps [2]:
- First decision: a step that depends on the editorial board and reviewers.
- Second review: a step that depends on the authors to revise and implement the received recommendations and comments.
- The time of backlog, i.e., time from acceptance to publication: the timeliness of this step varies according to whether the journal publishes papers online immediately after acceptance or you have to wait until the next issue is published. [31]

If rapid publication is your priority, it is important to check the average time required for each of these steps in the journal you are considering. It also might be helpful to see whether the journal allows you to track the review process. [6] Note that there is a correlation between a JIF and the time needed for its peer-review process. The higher the JIF is, the shorter the timeliness is expected to be. [32] Also note that some journals provide an optional service where authors can pay extra fees to hasten the peer-review and publication processes [33].

II. Acceptance Rate

Acceptance rate of a journal is often considered as an indicator for journal prestige or quality, i.e., the lower the acceptance rate is, the higher the quality of the journal is [7]. However, this might not be totally accurate. Reasons for that include the following:

- When calculating the acceptance rate, the denominator can vary according to the journal. Therefore, it might be useful to check the way your journal of choice calculates its acceptance rate. [34]
- Some journals favor certain types of articles. For example, some journals might accept review articles but reject case reports.
- There is a general bias in favor of English-speaking nations and prestigious institutions. Therefore, you might want to consider collaboration with authors from these backgrounds. [35]

Since the acceptance rate does not accurately reflect the journal quality, you should consider journals with a high acceptance rate if you aim for rapid publication as it will save you the time that might be consumed by multiple submissions.

To find a journal's acceptance rate, you might have to search multiple sources as this data may not be plainly stated on the journal website. These sources include the following:
- Journal editor reports
- Journal finding tools
- Metric pages within the journal
- *Journal* editor: you can contact the journal editor directly if this information is not available on their website [34]

III. Periodicity

The higher the periodicity of a journal is, the higher your chances of acceptance are because it means that the journal accepts more articles per year [2]. To get an idea of journal periodicity, you should be looking for the following:
- Number of volumes: How many years that journal has been in publication
- Number of issues: How frequently that journal publishes per year (semimonthly, monthly, bimonthly, quarterly, tri-annually, semiannually, annually).
- Number of articles published per year [36]

9.3.3.3 Publication Charges

To minimize publication charges, you need to consider two factors.

I. Type of Journal

In terms of publication fees, there are two main modes of publication (Table 9.4).

A third mode is hybrid journals which charge fees nearly twice as much as full OA ones. [37]

Table 9.4 A comparison between modes of publication

	Subscription journals	OA journals
Publication costs	Readers (subscribing individuals or institutions)	Authors
Exposure	Only subscribers can read your paper; hence, less audience and less citations	Your paper will be available for the general public; hence, more audience and more citations

However, there are exceptions to the previous rules:
- Not all OA journals charge fees. You can search DOAJ for no-fee journals in your field. [33]
- Some subscription journals charge authors for the following:
  - Publication processing charges
  - Exceeding word count
  - Exceeding limit allowed for number of figures, especially colored ones
  - Costs for printed pages [2, 38]
- Some journals offer reduction or complete wavering of charges in special cases where authors have scarce resources:
  - Authors from developing countries
  - Authors with no funding
  - Investigator-initiating studies (IISs) [7]

This variability and complexity of publication fees can make authors an easy prey for predatory journals [39]. Thus, it is essential that authors look for journals that offer transparency regarding charging details.

II. Funding

This will be covered in detail in the next chapter.

At the end of this chapter, you should remember the following:
- It is best if you prepare only a draft, pick the most suitable journals, and then write your manuscript.
- The visibility and credibility of your papers and, thus, your career advancement largely depend on your choice of journals.
- It is crucial to carefully study the scope of the journals you are considering.
- The eligibility of a journal must be thoroughly investigated as publishing in a predatory journal will kill the credibility of your research. The "Think. Check. Submit." checklist provides you with the points you need to check before submitting to any journal.
- Despite the controversy surrounding JIF, it is still the most popular indicator of the journal prestige.
- Subscription journals are journals that charge the readers, whether individuals or institutions, while OA journals are the ones that charge the authors. However, some journals offer reduction of charges in case of scarce resources.

References

1. Choosing the right journal for your research | Download your guide [Internet]. Author Services. [cited 2024 Nov 26]. Available from: https://authorservices.taylorandfrancis.com/resources/choosing-a-journal-free-guide/
2. Bahadoran Z, Mirmiran P, Kashfi K, Ghasemi A. Scientific publishing in biomedicine: how to choose a journal? Int J Endocrinol Metab. 2021;19(1)
3. Roldan-Valadez E, Salazar-Ruiz SY, Ibarra-Contreras R, Rios C. Current concepts on bibliometrics: a brief review about impact factor, Eigenfactor score, CiteScore, SCImago journal rank, source-normalised impact per paper, H-index, and alternative metrics. Ir J Med Sci. 1971). 2019 Aug 1;188:939–51.

4. Nordin MN, Khalid NF, Abd Ghani AH, Maidin SS, Abbas MS, Hamdzah NA, Mail S. I10 index and academics visibility. Int J Commun Networks Info Sec. 2024;16(3):159–65.
5. Schein M, Farndon JR, Fingerhut A. Why should a surgeon publish? A Surgeon's Guide to Writing and Publishing. 2001;1:1.
6. Thompson PJ. How to choose the right journal for your manuscript. Chest. 2007;132(3):1073–6.
7. Bavdekar SB, Save S. Choosing the right journal for a scientific paper. J Assoc Physicians India. 2015;63(6):56–8.
8. Masic I. How to search, write, prepare and publish the scientific papers in the biomedical Journals. AIM. 2011;19:68–9.
9. Meyer HS, Durning SJ, Sklar DP, Maggio LA. Making the first cut: an analysis of academic medicine editors' reasons for not sending manuscripts out for external peer review. Acad Med. 2018;93(3):464–70.
10. Jawaid SA, Jawaid M. Common reasons for not accepting manuscripts for further processing after editor's triage and initial screening. Pakistan J Med Sci. 2019 Jan;35(1):1.
11. Cals JW, Kotz D. Effective writing and publishing scientific papers, part X: choice of journal. J Clin Epidemiol. 2014;67(1):3.
12. Danh NN, Tran TR, Tien TN. Thi TT. Manuscript Matcher: A Tool Finding Best J.
13. Rison RA, Shepphird JK, Kidd MR. How to choose the best journal for your case report. J Med Case Rep. 2017 Dec;11:1–9.
14. Journal Recommendation and Selection Service for Publication I Wiley Editing Services [Internet]. [cited 2024 Nov 26]. Available from: https://wileyeditingservices.com/en/article-preparation/journal-recommendation?utm_source=as&utm_medium=referral&utm_term=findjournal&utm_content=wesjr&utm_campaign=prodops
15. Cobey KD, Lalu MM, Skidmore B, Ahmadzai N, Grudniewicz A, Moher D. What is a predatory journal? A scoping review. F1000Research. 2018;7:7.
16. Elmore SA, Weston EH. Predatory journals: what they are and how to avoid them. Toxicol Pathol. 2020;48(4):607–10.
17. Choudhary M, Kurien N. Predatory journals: a threat to evidence-based science. Indian J Health Sci Biomed Res kleu. 2019;12(1):12–4.
18. Ferris LE, Winker MA. Ethical issues in publishing in predatory journals. Biochem Med. 2017;27(2):279–84.
19. Gurnani B, Kaur K. Predatory journals: the dark side of publications. Indian J Ophthalmol. 2022;70(8):3144–5.
20. da Silva JA, Tsigaris P. What value do journal whitelists and blacklists have in academia? J Acad Librariansh. 2018;44(6):781–92.
21. Strinzel M, Severin A, Milzow K, Egger M. Blacklists and whitelists to tackle predatory publishing: a cross-sectional comparison and thematic analysis. MBio. 2019;10(3):10–128.
22. Marchitelli A, Galimberti P, Bollini A, Mitchell D. Helping journals to improve their publishing standards: a data analysis of DOAJ new criteria effects. JLIS it. 2017;8(1):1.
23. Greenwood DC. Reliability of journal impact factor rankings. BMC Med Res Methodol. 2007;7:1–6.
24. Togia A, Tsigilis N. Impact factor and education journals: a critical examination and analysis. Int J Educ Res. 2006;45(6):362–79.
25. Bonato F. Journal metrics. Aerospace Medicine and Human Performance 2016 Jan 1;87(1):1-, 1.
26. Villaseñor-Almaraz M, Islas-Serrano J, Murata C, Roldan-Valadez E. Impact factor correlations with Scimago journal rank, source normalized impact per paper, Eigenfactor score, and the CiteScore in radiology, nuclear medicine & medical imaging journals. Radiol Med. 2019;1(124):495–504.
27. Pendlebury DA. The use and misuse of journal metrics and other citation indicators. Arch Immunol Ther Exp. 2009 Feb;57:1–1.
28. Huh S. How to add a journal to the international databases, science citation index expanded and MEDLINE. Arch Plast Surg. 2016;43(06):487–90.
29. Journals • Think. Check. Submit [Internet] Think Check Submit [cited 2024 Nov 26]. Available from: https://thinkchecksubmit.org/journals/

30. Knight LV, Steinbach TA. Selecting an appropriate publication outlet: a comprehensive model of journal selection criteria for researchers in a broad range of academic disciplines. Int J Dr Stud. 2008 Jan;1:3.
31. Welch SJ. Selecting the right journal for your submission. J Thorac Dis. 2012;4(3):336.
32. Huisman J, Smits J. Duration and quality of the peer review process: the author's perspective. Scientometrics. 2017 Oct;113(1):633–50.
33. Panter M. Understanding submission and publication fees. American J Experts. https://www.aje.com/en/arc/understanding-submission-and-publication-fees/, [dostęp: 09.02. 2018]. 2016.
34. Services EA. Journal Acceptance Rates: Everything You Need to Know [Internet]. Elsevier Author Services – Articles. 2022 [cited 2024 Nov 26]. Available from: https://scientific-publishing.webshop.elsevier.com/publication-process/journal-acceptance-rates/
35. Ross JS, Gross CP, Desai MM, Hong Y, Grant AO, Daniels SR, Hachinski VC, Gibbons RJ, Gardner TJ, Krumholz HM. Effect of blinded peer review on abstract acceptance. JAMA. 2006;295(14):1675–80.
36. Singh A, Singh S, Mercy P, Singh AK, Singh D, Singh M, Singh P. Art of publication and selection of journal. Indian Dermatol Online J. 2014;5(1):4–6.
37. Björk BC, Solomon D. How research funders can finance APCs in full OA and hybrid journals. Learned Publishing. 2014;27(2):93–103.
38. García JA, Rodriguez-Sánchez R, Fdez-Valdivia J, Chamorro-Padial J. The author's ignorance on the publication fees is a source of power for publishers. Scientometrics. 2019;121:1435–45.
39. Hopf H, Krief A, Mehta G, Matlin SA. Fake science and the knowledge crisis: ignorance can be fatal. R Soc Open Sci. 2019;6(5):190161.

How to Properly Submit Your Paper

10

Abdullah Olimy, Mohamed Ahmed Ali, Islam Mohammad Shehata, Omar Viswanath, and Robert Ravinsky

> *It is not how you start the race or where you are during the race—it is how you cross the finish line that will matter.*
>
> —Robert D. Hales

10.1 Preparing for Submission

This stage comes after completing your final version of your manuscript and selecting the suitable journal to which you will submit the paper. This is where you will start to format your manuscript files according to the journal's guidelines, create an account on the journal's submission portal, and submit your manuscript.

A. Olimy (✉)
Faculty of Medicine, Menoufia University, Menoufia, Egypt

M. A. Ali
Qena Faculty of Medicine, South Valley University, Qena, Egypt

Medical Research Group of Egypt (MRGE), Negida Academy, Arlington, MA, USA

I. M. Shehata
Department of Anesthesiology, Faculty of Medicine, Ain Shams University Cairo, Cairo, Egypt
e-mail: Islam.shehata@med.asu.edu.eg

O. Viswanath
Department of Anesthesiology, Creighton University School of Medicine, Phoenix, AZ, USA

Mountain View Headache and Spine Institute, Phoenix, AZ, USA

R. Ravinsky
Department of Orthopaedics and Physical Medicine, Medical University of South Carolina, Charleston, SC, USA
e-mail: rar277@mail.harvard.edu

© The Author(s), under exclusive license to Springer Nature Switzerland AG 2025
I. M. Shehata, O. Viswanath (eds.), *How to Successfully Publish a Manuscript*,
https://doi.org/10.1007/978-3-031-92538-2_10

10.1.1 Checking Journal Submission Guidelines

- You must carefully read the submission guidelines of the journal. It may be listed under the name of **Submission Guidelines**, **Information for Authors**, etc. It is composed of a list of instructions that should be applied to the manuscript, and they differ from one journal to another and from one type of manuscript to another.
- Most journals ask to upload your manuscript data in separate files:
  - Title page
  - Manuscript body (title, abstract, full manuscript, and references) without any author data
  - Cover letter
  - Any figures, tables, and artwork may be collected in one file or each one uploaded separately one by one
- However, others ask to combine all these files in just one file. As you may see "Title page" and "Cover letter" for the first time, they will be further discussed in detail.

10.1.2 Account Creation

To submit your manuscript, you must create an account on the journal's submission portal, by following a couple of steps.

- **Example**: From the website of *Nature Medicine* journal (nature.com/nm/). Some steps may differ from one journal to another.
- A. You first go to the journal website and then click on (submit, submit a manuscript, or submit an article, etc.) (Fig. 10.1).

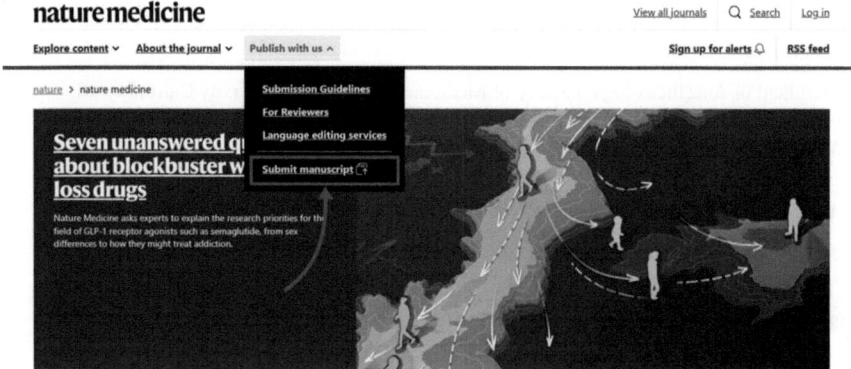

Fig. 10.1 Home page of nature medicine journal

B. It will refer you to the submission portal, click on "Register for an account" to start creating your account (Fig. 10.2).
C. Here, you enter your "Last Name" and "Primary Email Address" and then click "Continue" (Fig. 10.3).
D. Here, you must enter the mandatory fields "*", which include First Name, Last Name, Username, Password, and Confirm Password. Read the "**Terms and Conditions**" and check their box. Make sure to keep your **Username** and **Passwords** as you will need them to log in your account. Then, press the "**Register**" button (Fig. 10.4).

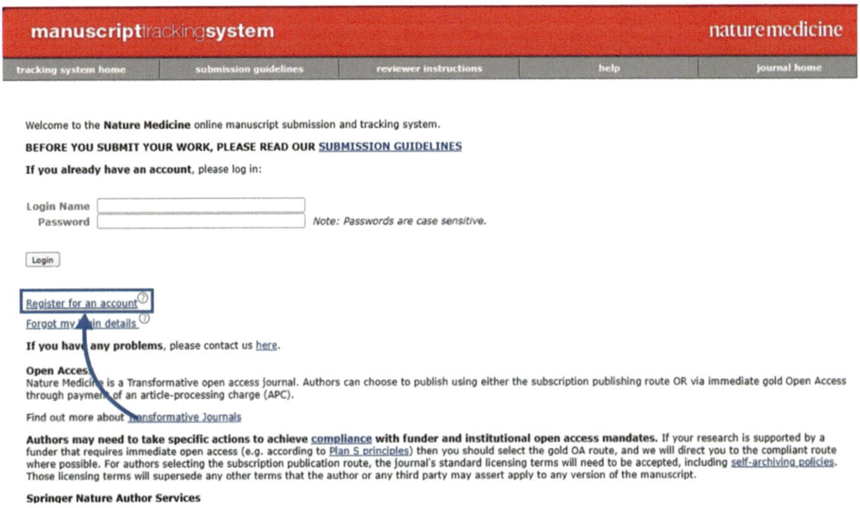

Fig. 10.2 Login page of the submission portal of nature medicine journal

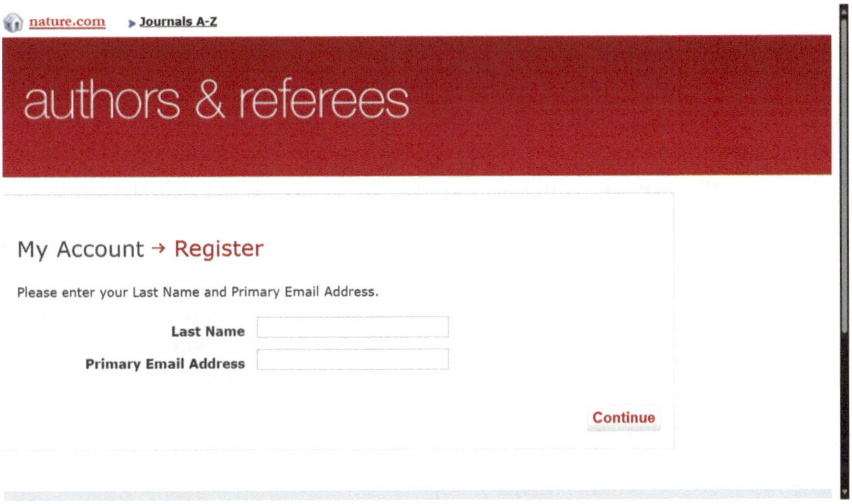

Fig. 10.3 Registration page of the submission portal of nature medicine journal

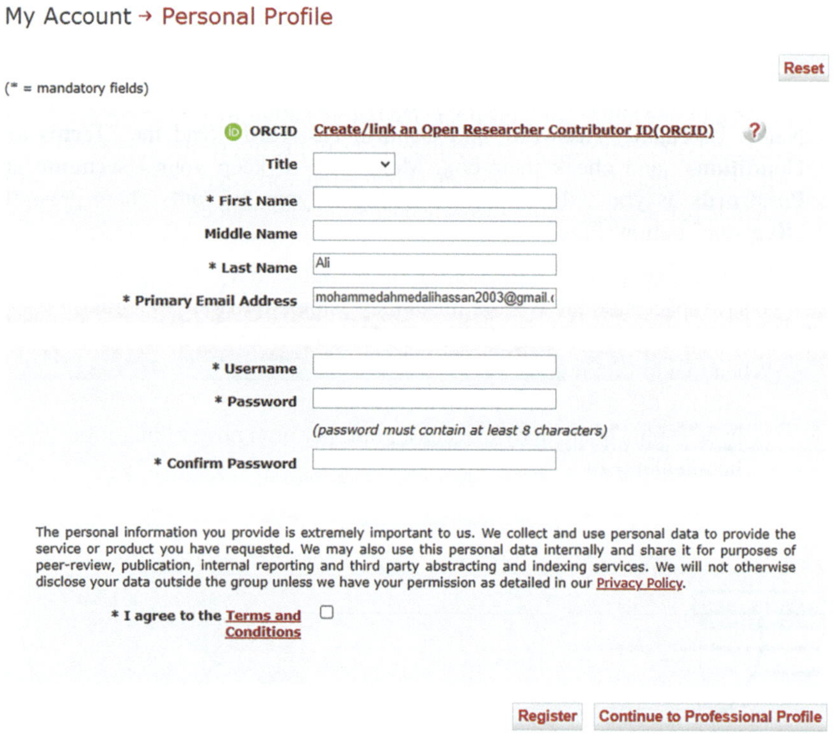

Fig. 10.4 Personal profile data for nature medicine journal

E. Now, you have created your profile and can upload your manuscript. It is optional if you want to add more information about you in the "**Professional profile**" tab. After adding the information you want, click "**Save**" (Fig. 10.5).

10.1.3 Uploading Files

After creating your account, you can start your submission process and upload your files through the journal's submission portal. As previously mentioned (1. **Checking Journal Submission Guidelines**), you must submit your files either separately or in one file, which will be according to the journal's guidelines. Also, they may ask for certain file types like Word or PDF, so make sure that your files are in the required format.

- **Example**: Also, from the website of *Nature Medicine* journal (nature.com/nm/).
A. The first step is the same as you did in "**1.2 Account creation.**"
B. You log in with your Login Name (Username) and Password and then click "Login" (Fig. 10.6).
C. Here, you find your homepage; click "**Submit a new manuscrip**t," so you can start submitting the manuscript (Fig. 10.7).

10 How to Properly Submit Your Paper   141

Fig. 10.5 Professional profile data for nature medicine journal

Here, according to your study type, you will be asked to upload the files. You should be able to easily follow the guidelines of the journals and upload all the required files.

10.1.4 Authorship Verification

Most journals send verification emails to all listed authors to their respective emails which the corresponding author added through the submission process. Some journals may ask for creating an account as a verification step or click on a verification link, while others only ask the author to contact if there is something wrong with their data or if they are listed as authors by mistake on that particular manuscript.

Also, there are other journals that only send an email to the corresponding author, and he/she is responsible for making sure that all authors' data are right and that all listed authors are the real authors of the paper.

– **Example**: This is a verification email sent from the "**BMC Cardiovascular Disorders**" journal, with sensitive information masked (Fig. 10.8).

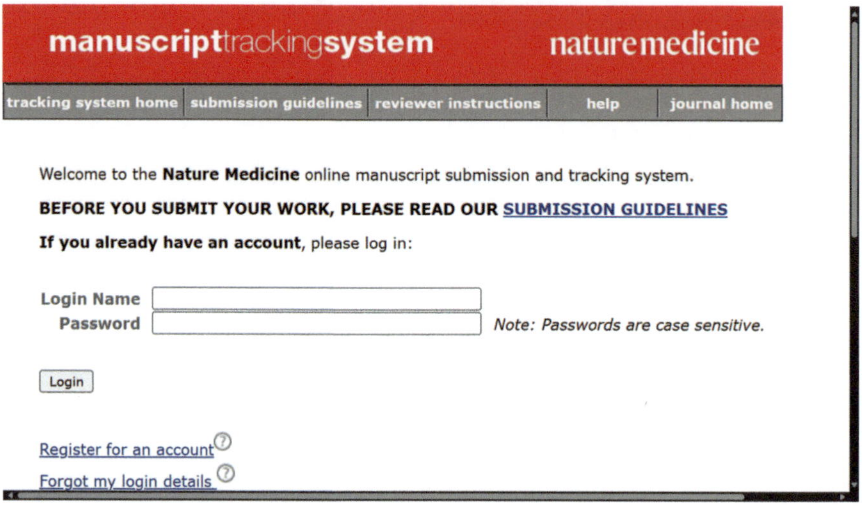

Fig. 10.6 Login page of the submission portal of Nature Medicine journal

Fig. 10.7 Submission portal of nature medicine journal

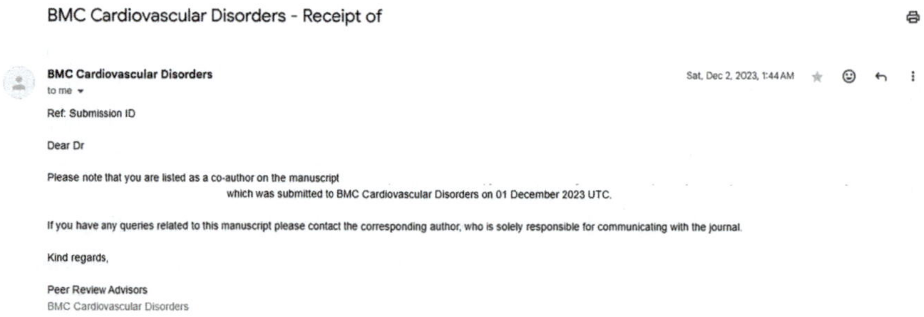

Fig. 10.8 Email of authorship verification

10.2 Preparing the Title Page

10.2.1 Definition

- The title page is the first part of the submitted manuscript. It has two types:
  - **Blinded title page**: It only has the manuscript title, and it is sent with the complete blinded manuscript for reviews to ensure a peer-review process unbiased with masking authors who have written the paper [1].
  - **Unblinded (complete) type**: Shows manuscript's title, authors' names, their affiliations, and the full data of the corresponding author.

10.2.2 Structure

1. **Manuscript Title**: Title is the first part the Editor and Reviewers see (refer to the title chapter).
2. **Keywords and Running Title**
3. **Authors' Data:**
   You should provide the data for all the authors who took part in the manuscript.
   - Essential data includes:
     1. *Full Author Name*
     2. *Email (preferable to be an institutional email)*
     3. *The Affiliation*
     4. *Contact Number*
     5. *Postal Address*
     6. *Postal (Zip) Code*
   - **Affiliation:** This is the name of the Department, Institution, City, and Country where the author was working while doing the manuscript.
     - **Example**: Qena Faculty of Medicine, South Valley University, Qena, Egypt.
   - **NB**: Some journals ask for Author's Open Researcher and Contributor ID (**ORCID**).
   - **ORCID**: This is (as mentioned in their official website) a free service which enables the researchers to make a persistent identifier (**PID**) formed of 16 numbers.
     - **Example**: **0000-0001-8142-8073** which is linked to a profile on their websites where they can add all their research publications, awards, education, and any professional activities. Also, it can be used while signing in to different journals' websites.
     - **How to get yours**?
       You can visit the official channel of **ORCID** on Vimeo for different tutorials about using it (https://vimeo.com/orcidvideos). This showcase is very useful for a guide for those who want to start using **ORCID** (https://vimeo.com/showcase/4268215).
4. **Corresponding Author's Data:**

- **Role:** The **Corresponding Author** is the author responsible for:
  - *Contacting the journal.*
  - *Preparing the manuscript for submission.*
  - *Submitting the manuscript.*
  - *Corresponding with the journal throughout the peer-review process.*
  - *Receiving and replying to reviewers' comments.*
  - *Signing any required documents from the journal which are related to the license of the paper.*
- **Who?**
  It is recommended that the Corresponding Author typically be the first or last author as they are the primary author who has been coordinating the team.
  NB: While most journals do not ask for much data about other authors, detailed data is asked from the corresponding author.
- **Required information:**
  - Corresponding author's data includes name, affiliation, e-mail, ORCID, phone number, and sometimes the address.
- **Example:** (**Fig. 10.9**).
- **Example of a full title page:** (Fig. 10.10).

10.2.3 Other Sections

10.2.3.1 Conflict of Interest Statement
You should declare if there are any competing interests that may interfere with the integrity of the manuscript, such as:

- Financial support.
- Nonfinancial (personal, ideological, academic, etc.)
- Relationship with commercial organizations (pharmaceutical companies or others).
- If there are no conflicts of interest, you also should write this clearly. It may be named "Competing interests" in some journals; so, you should read the submission guidelines of the journal and know what they need to be addressed.

*Corresponding author:

Mohamed Ahmed Ali

Affiliation: Qena Faculty of Medicine, South Valley University, Qena, Egypt.

mohammedahmedalihassan2003@gmail.com

https://orcid.org/0000-0001-8142-8073

Address: AR Raiseyah, Nag Hammadi, Qena, Egypt.

Fig. 10.9 Example of the data of the corresponding author

Title of the Manuscript:

Running title:
Keywords:
Names of the Authors:

AuthorA[1], AhuthorB[2], AuthorC[3], AuthorD[4], AuthorE[5], ...etc.

Authors' Data*:

1-
2-
3-
4-
5-

Corresponding Author*:
- Name.
- Affilation.
- Email.
- ORCID.
- Address.
- Postal (Zip) Code.

Fig. 10.10 Title page example

Competing interests

All financial and non-financial competing interests must be declared in this section.

See our editorial policies for a full explanation of competing interests. If you are unsure whether you or any of your co-authors have a competing interest please contact the editorial office.

Please use the authors initials to refer to each authors' competing interests in this section.

If you do not have any competing interests, please state "The authors declare that they have no competing interests" in this section.

Fig. 10.11 Conflicts of interest example

- Example: from "BMC Cardiovascular Disorders" journal. For full details, click link (https://www.biomedcentral.com/getpublished/editorial-policies#competing+interests) (Fig. 10.11).

Fig. 10.12 Word count from Microsoft Word

10.2.3.2 Acknowledgments (Not Essential)

Here, you must mention any contributions made by individuals or institutions that took part in the manuscript that do not reach the level of authorship.

The International Committee of Medical Journal Editors (ICMJE) advises that authorship must be based on four criteria:

1. "Substantial contributions to the conception or design of the work; or the acquisition, analysis, or interpretation of data for the work
2. Drafting the work or reviewing it critically for important intellectual content.
3. Final approval of the version to be published.
4. Agreement to be accountable for all aspects of the work in ensuring that questions related to the accuracy or integrity of any part of the work are appropriately investigated and resolved." (https://www.icmje.org/recommendations/browse/roles-and-responsibilities/defining-the-role-of-authors-and-contributors.html).

Therefore, any contributions that did not fulfill these criteria should be listed in the Acknowledgments section.

10.2.3.3 Word Count

Some journals ask for a minimum or maximum number of words, so you have to adhere to these limitations and mention the word number of your title page, cover letter, or whole manuscript that are mentioned in the Journal Guidelines. You can easily know it from Microsoft Word as it appears in the lowest ribbon beside the page number.

Example: (Fig. 10.12).

10.3 Cover Letter

An impactful cover letter is your first step to validate the impact and credibility of your work.

10.3.1 Definition [2]

It is a professional document submitted alongside the manuscript to the journal of interest introducing your work. It is the responsibility of the corresponding author to finely craft a decent cover letter that introduces your work to the editor-in-chief explaining why they should publish your manuscript to their journal.

From Whom? To Whom?
Corresponding author Editor-in-chief.

10.3.2 Importance

The cover letter is your chance to outline why your paper is suitable for the journal, emphasize the importance of your findings, and address any special requirements [2, 3]. (Fig. 10.13).

10.3.3 General Theme of a Decent Cover Letter

Writing the cover letter is a crucial part of your process of publication. It is your first pitch to make a decent lasting impression about who you are and your work impact (Fig. 10.14).

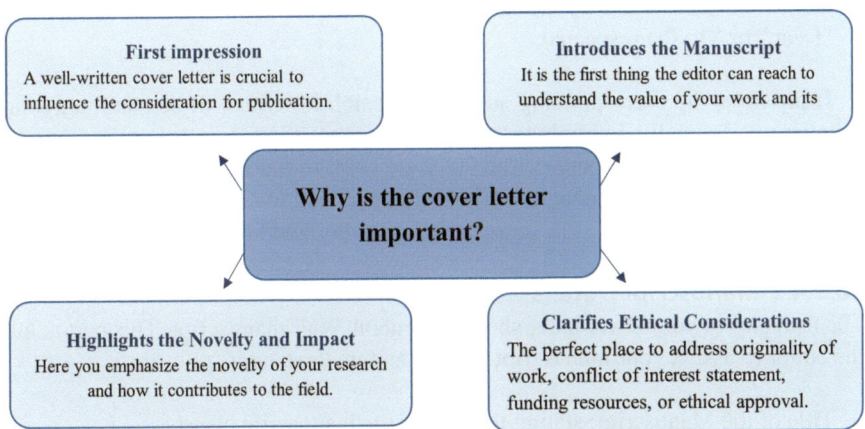

Fig. 10.13 Importance of cover letter

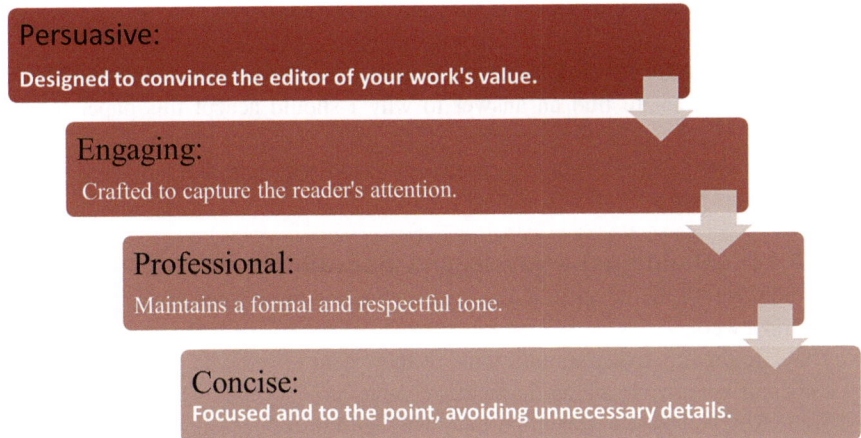

Fig. 10.14 Characteristics of a cover letter

Your cover letThe letter should be a mix of a professional introduction to your work and a strong formal persuasive argument for your finding validation. In the end, you expect your cover letter to tell the journal's editorial board that your manuscript has well-founded outcomes that will be important for their readers to learn, which ultimately is the best fit for their journal.

10.3.4 Structure

10.3.4.1 Addressing the Editor with Professionalism
- The cover letter is intended to engage the attention of the journal's editorial board, specifically the editor-in-chief.
- You should start by addressing the editor by name if it is available. But in general, if it is not available, you begin with a general salutation such as "Honorable," "Dear," or "To the respected."

If possible, the corresponding author can search for who will be handling your submission. Typically, journals list the editor responsible for each section or category of research on their site. Directing your work to the right individual reflects your interest and commitment in addition to a personal touch. But the reflection of this step depends on who is the corresponding author and his achievements.

10.3.4.2 Manuscript Details
The next part of your cover letter should talk about your manuscript. This part helps the editor to assess your manuscript in brief and understand its relevance.

- Title of the Manuscript: stating the title of the manuscript clearly.
- Type of Submission: classify your work whether it is a systematic review, clinical trial, case report, meta-analysis, etc.
- Originality: you should confirm that this manuscript is your original work, not published before or considered to be published elsewhere.

10.3.4.3 Rationale for Submission
- The editors want to find an answer to why I should accept this paper to the journal.
- You start justifying why your research fits this journal scope and why they should publish this paper.

10.3.4.4 Highlighting the Novelty and Innovation
- Demonstrating the novelty of your research is essential to convince the editors and reviewers.
- They are reading many submissions, so they want to see what is new and what can contribute to advancing the field.
- Show why your research is important, whether it introduces new methodologies, reveals previously unexplored data, or challenges existing paradigms.

10.3.4.5 Ethical Conformity and Disclosure of Conflicts of Interest
– Provide a brief statement confirming your adherence to ethical guidelines, especially if your study involves the participation of living subjects, whether it is human or animal.
– Ethical guidelines are non-negotiable. Many journals will reject manuscripts for unclear ethical statements without a thought.
– Additionally, disclosure of any potential conflicts of interest (on behalf of all authors) clearly is essential to maintaining the integrity of the process.

10.3.4.6 Closing Remarks
– End your cover letter with gratitude and openness.
– Thank the editor and provide your contact information for follow-up, any questions, or clarifications.
– A professional polite closing ensures that the final impression about your submission stays positive.

10.3.4.7 Example (Fig. 10.15)

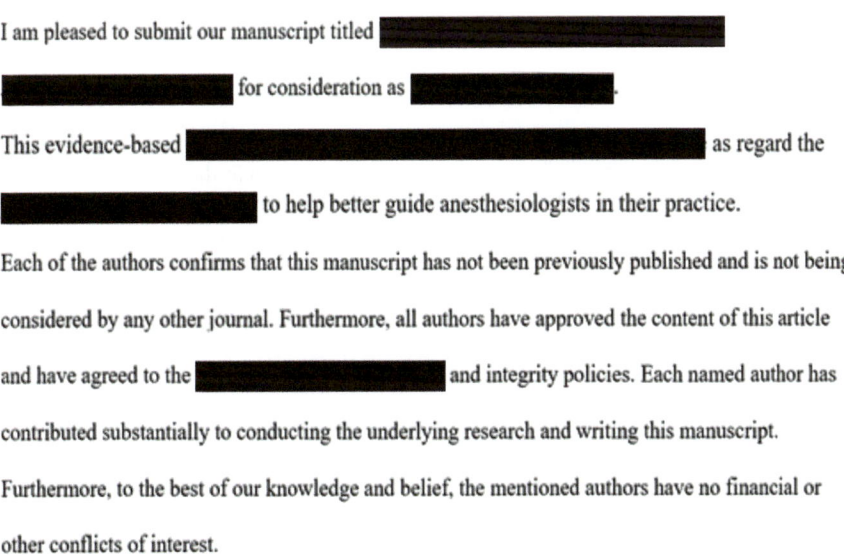

Fig. 10.15 Cover letter example

10.4 Post-Acceptance Phase: In Press and Publication

It is time to celebrate, start promoting your hard work.

10.4.1 Definition

It is the transitional period after the manuscript is accepted for publishing, encompassing the final preparation including (assigning Digital Object Identifier (DOI), proofreading, typesetting), and early online access ("in press") [4] (Fig. 10.16).

10.4.2 What to Expect After Acceptance? [4]

- Following the editorial process after submission, your manuscript is sent to final production for typesetting.
- The journal asks for some secure steps which are done by the corresponding author, including:
  - Confirming affiliation (institution or university).
  - Choose whether to publish as an open access or non-open access (subscription with fees).
  - Determine whether there is an agreement from your institution to cover the article processing charge (**APC**).

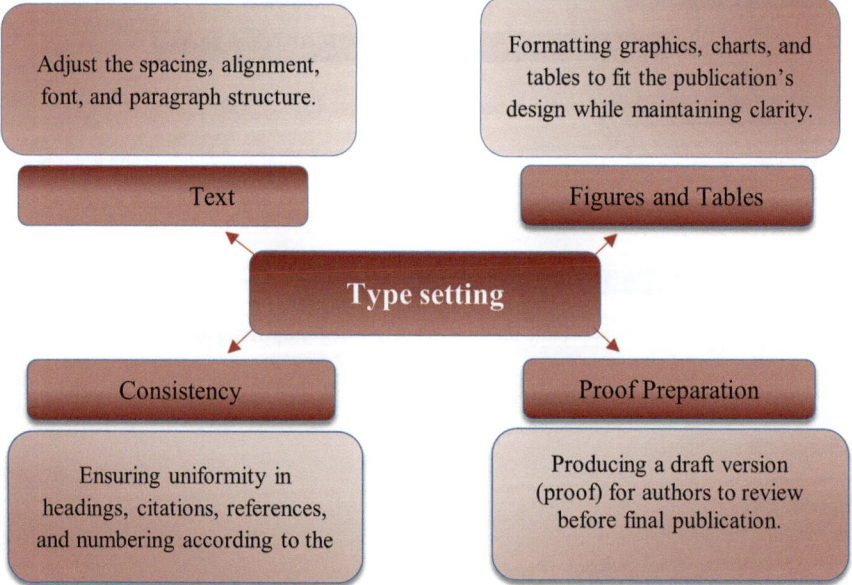

Fig. 10.16 Characteristics of typesetting

- Sign the suitable publishing agreement.
- Arrange payments for associated publication costs.

NB: Your publication only proceeds to publication after completion of these steps, starting to be assigned with DOI, "**In Press**" issued, and finally fully published.

10.4.3 Assigning Digital Object Identifier (DOI) [5]

- Digital object identifier (DOI): is a stable link to the published content. Usually assigned by DOI registration agencies like Crossref.
- It guarantees that scholarly works have the correct and same citation links and metadata registration.
- **Permanent Access:** Unlike traditional URLs, DOI is a unique and permanent digital identifier link.
- **Citable Link:** DOI is commonly used in scholarly citations to ensure that readers can always find the original regardless of the URL changes.

10.4.4 In Press: What Does It Mean? [6]

"**In Press**" mainly refers to the period after acceptance and before the paper is assigned to a journal issue. During this stage, the paper will not be assigned with date, volume, issue, number of pages, or even in some cases a DOI.

- **Early Access and Indexing**: Your paper will be accessible by all researchers and readers, as it will be indexed in databases like PubMed, allowing potential citations and helping you promote your work earlier.
- **Proofs Review**: You will receive the "proofs" of your article following acceptance. This is your final opportunity to fix any possible mistakes.

10.4.5 Present Your Work!

- The publication is only the beginning of your journey; it is not the end!
- In the academic and clinical world, researchers find that sharing their results is beneficial; Discussing your latest discoveries, intriguing theories, or uncommon situations with others offers a priceless chance to hone your research concepts advance your scientific thinking, find exciting research questions, and perhaps form new partnerships.
- These conversations not only foster intellectual growth but also ensure that your colleagues stay up to date on the latest developments, enabling them to apply these realizations to their work.
How to get involved in these discussions?

- **Conferences and Seminars:** Giving a presentation at pertinent clinical or academic gatherings. Emphasize your findings to increase the effect of your work.
- **Academic Profile:** Update your information on scholarly websites such as Google Scholar, ResearchGate, and ORCID.
- **Making Use of Online Platforms:** Social media is a great way to reach the proper audience with your most current publications. To choose the proper audience, you must choose professional platforms to promote your work like LinkedIn and Twitter.

References

1. Peh WCG, Ng KH. Title and title page. Singapore Med J. 2008;49:607–608; quiz 609.
2. Cover letters | Springer — International Publisher. https://www.springer.com/gp/authors-editors/authorandreviewertutorials/submitting-to-a-journal-and-peer-review/cover-letters/10285574.
3. Vale M. How to write a cover letter for your manuscript | Elsevier. Elsevier Author Services – Articles. https://scientific-publishing.webshop.elsevier.com/publication-process/how-to-write-a-cover-letter-for-a-manuscript/; 2024.
4. Next steps for publishing your article: What to expect after acceptance. Springer Support https://support.springer.com/en/support/solutions/articles/6000084585-next-steps-for-publishing-your-article-what-to-expect-after-acceptance.
5. Rosa-Clark. Creating and managing DOIs. Crossref https://www.crossref.org/documentation/register-maintain-records/creating-and-managing-dois/.
6. Lamb, A. Guides: APA 7th edition – University of Lincoln: advance online publications or articles in press. https://guides.library.lincoln.ac.uk/c.php?g=683973&p=4882454.

How to Raise Funds for Publication Fees? 11

Salwa Khaled Ahmed, Eman Hamdy Oweiss, Islam Mohammad Shehata, Omar Viswanath, and Alberto Pasqualucci

> *If opportunity doesn't knock, build a door.*
>
> —Milton Berle

11.1 Definition

Funds are defined as "an amount of money that has been saved or has been made available for a particular purpose." [1] When it comes to publishing, funds refer to the cost of the publication process.

This rising cost of publication has been a limiting factor for authors, especially those without financial support.

S. K. Ahmed (✉)
Faculty of Medicine, Cairo University, Cairo, Egypt
e-mail: salwa_k_ahmed@students.kasralainy.edu.eg

E. H. Oweiss
Faculty of Medicine, Ain Shams University, Cairo, Egypt
e-mail: 170950@med.asu.edu.eg

I. M. Shehata
Department of Anesthesiology, Faculty of Medicine, Ain Shams University Cairo, Cairo, Egypt
e-mail: islam.shehata@med.asu.edu.eg

O. Viswanath
Department of Anesthesiology, Creighton University School of Medicine, Phoenix, AZ, USA

Mountain View Headache and Spine Institute, Phoenix, AZ, USA

A. Pasqualucci
Anesthesia and ICU, Department of Medicine and Surgery, University of Perugia, Perugia, Italy
e-mail: alberto.pasqualucci@unipg.it

© The Author(s), under exclusive license to Springer Nature Switzerland AG 2025
I. M. Shehata, O. Viswanath (eds.), *How to Successfully Publish a Manuscript*, https://doi.org/10.1007/978-3-031-92538-2_11

11.2 What Are Publication Fees? [2]

There is variability and complexity of publication fees which can make authors easy targets for predatory journals [3]. Therefore, authors must look for journals that offer transparency regarding charging details.

Before Peer-review
- **Membership fees**: For subscription journals that allow subscribers to publish.
- **Submission fees**: Paid to cover the cost of submission and are usually nonrefundable even if the manuscript was rejected by the journal.

After Acceptance
- **Article processing charges (APCs)**: Most encountered when it comes to open-access publishing [4]
  - They can vary between a few hundred dollars to thousands according to the quality and profile of the journal.
  - APCs cover editing, proofreading, plagiarism checking, archiving, marketing, suitable indexing, website maintenance, and customer service.
- **Printing charges:** For the printed version of the manuscript, they are charged per number of pages and/or color figures.

11.3 Understanding Different Open-Access (OA) Models [5]

Models of OA (see Table 11.1 follow a color-naming system; most recognizable are "green," "gold," "diamond," and "hybrid." An understanding of OA models is crucial, especially when applying for funding as different funding organizations have OA policies, which can limit the models or journals the author wishes to publish in.

11.4 What Are the Sources of Funding Available?

The most common way to cover APCs is out-of-pocket; however, there are many sources you should look for first.

- **Author waiver or discounts**: Offered by some journals and automatically apply to authors in certain geographical locations (developing countries); search for your country on the journal's website. You may find a list of the eligible countries either for a waiver (complete deduction of fees) or discount (a percent deduction of fees).
- **Publishing agreements**: Check for any preexisting agreements between the publisher and your institution.

Table 11.1 Open-access models

Green OA	A self-archiving model where the author can deposit their manuscript in an institutional repository or even a personal website; without paying APCs. ☐ The version of the manuscript is up to the publisher or funder. ☐ There is typically an embargo period meaning the article is not made open-access until a certain period set by the publisher has passed. ☐ You can search for open-access repositories in your country at OpenDOAR.
Gold OA	A model where the manuscript is immediately and freely available upon publishing. ☐ Requires a payment of APCs which can be paid by the author but is typically funded by the author's affiliated institution. ☐ The author retains copyright under a Creative Commons license.
Hybrid OA	Subscription-based journals that allow authors to publish open-access by paying APCs but can cost more than gold OA. ☐ The manuscript is immediately available, however only articles that were made open-access are free even for non-subscribers.
Diamond OA	A model where the author can publish without fees and the publisher is funded by an external source such as institutions, governmental grants, scientific societies, etc. ☐ To find such journals you can use the "without fees" search filter option on DOAJ: Directory of Open Access Journals.

Examples:
- The Egyptian Knowledge Bank collaboration with Springer Nature covers the cost of APCs for Egyptian authors; you can view the list of journals included on Springer Nature's website [6].
- Projekt DEAL launched by the Alliance of Science Organizations in Germany has agreements with major publishers (Wiley, Springer Nature, and Elsevier) to facilitate open-access publishing for authors in Germany by offering discounts on APCs [7].
- Elsevier and Manipal Academy of Higher Education (MAHE) in India have an agreement to cover APCs for eligible authors [8].
- Wiley and Joint Information Systems Committee (JISC) in the UK cover APCs for eligible authors [9].

- **Institutional support**: Some universities/institutions have funds to cover publication fees although some limitations exist on the amount that can be covered or specific requirements. You can access this fund through your institute or a **co-author's** institute.
- **Professional societies**: Cover APCs if you publish with their affiliated journals. Example: American Medical Association (AMA) provides discounts for members wishing to publish with their journal.
- **Research grant**: You can include APCs as part of the grant budget. This can be justified as a necessary cost associated with the dissemination of the research results.

Example: The National Institutes of Health (NIH) RO1 grants are considered the gold standard. The NIH publishes a weekly table of contents and new funding opportunities list to its email LISTSERV.

- 💰 **Funding organization:** They are third party organizations that can provide funding to cover APCs.

11.5 How to Find the Funding Organization?

- Start by looking for organizations directly interested in your area of research.
- Look out for what topics of research got funding the previous year; this allows you to get an idea of what they are looking for (see Table 11.2).

Table 11.2 Funding organizations in some countries)

Country	Organization/ institution	Funding	Limitation
Australia	National Health and Medical Research Council (NHMRC)	NHMRC grant funds cover APCs [10]	Eligibility criteria apply to different grant programs
Australia	Australian Research Council (ARC)	ARC grant funds cover APCs	Grants are handed to organizations (usually an Australian university) and not individual researchers [11]
Canada	Canadian Institutes of Health Research (CIHR)	CIHR grant funds cover APCs	Researchers based in an eligible Canadian institution
Europe	European Research Council (ERC)	ERC grant funds cover APCs.	Researchers based in an EU member state or associated countries
Germany	German Research Foundation (DFG)	Offers publication grants that fund electronic book publications and open-access *Only institutions can apply for "open-access publication funding"; however, individual researchers can request funding for APCs as part of their DFG grant*	Researchers based in Germany Publication grant is only awarded to works: °°Of exceptional scientific importance °°Basic material is made available [12]
Ireland	Science Foundation Ireland (SFI)	SFI grant funds cover APCs	Researchers based in Ireland
Japan	Japan Society for the Promotion of Science (JSPS)	Offers a KAKENHI grant program that covers publication fees [13]	Researchers based in Japan

(continued)

Table 11.2 (continued)

Country	Organization/institution	Funding	Limitation
Qatar	Qatar National Library	Funding through Open-access fund application (Up to 3000 USD) *Criteria include journals registered in DOAJ and authors are limited to 3 articles per year* Agreements [14]	Affiliation with a Qatar-based nonprofit institution
UK	Wellcome Trust Fund	Wellcome grant funds cover APCs	Fully OA Journals: °°Indexed by DOAJ °°Have agreements to make the Version of Record shareable with Europe PMC
USA	Bill & Melinda Gates Foundation	Gates grant funds cover APCs	Journals that align with the foundation's OA policy *To look for compliant journals you can use the journal checker tool* [15]

11.6 Proposal Writing

The key to securing a fund is the ability to write a compelling grant proposal that piques the organization's interest (see Table 11.3).

Writing rules
Elements of a Proposal
The following are basic components that should be present in any proposal; however, make sure to **check with the guidelines** for any additional elements.

1. Title Page
    This should contain the following:

> [Title of your manuscript]
> [Corresponding author's name]
> [His/Her position]
> [His/Her affiliation]
> [Address of your institution]
> [E-mail address]
> [Phone number].

Table 11.3 Dos and don'ts in proposal writing

DO'S	DON'TS
☑ **Be straightforward**: Use clear simple language.	☒ **Be vague**: Back up your proposal with your previous work.
☑ **Start early**: To have enough time for proofreading and making adjustments.	☒ **Ignore the guidelines**: they differ according to each organization.
☑ **Sell your idea**: Tailor your proposal in a way that aligns with the funding organization's views.	☒ **Miss Information**: attach all the needed documents according to the guidelines.
	☒ **Lack context**: highlight the importance of your article being published.

2. Executive Summary Scenarios:

I am writing to formally request funds to support the publication of my manuscript [title of manuscript] which has been accepted for publishing in [journal name] and is expected to be published on [date]. I am requesting [X amount] which will cover the article processing charges.

- In case a journal still has not been selected, you can mention which journals you anticipate publishing in and the range of APCs.

I'm writing to formally request funds to support the publication of my manuscript [manuscript title] which I anticipate publishing in [list the journal names]. The cost of article processing charges for these journals ranges between [X amount to X amount], therefore I'm requesting [X amount] in funding.

- Followed by a brief explanation of the purpose of your research.

My manuscript addresses [state the purpose] and I believe its publication in [journal name] would provide a valuable contribution to [describe the relevance to solving the problem at hand]. I appreciate [organization's name] for its efforts to be part of the solution.

- If your research was funded by any program/organization, it should be acknowledged here.

This research received funding in whole/in part from [organization's name]

3. **Background and Significance**
   Highlight the impact of your research getting published, what makes it different, and its importance to the scientific community
   - Make sure to include the gap of existing literature review and the results of preliminary work to back up your claims further.
4. **Publication Details**.

> **Name of journal (if available)**
> **Explain the choice of journal**
> **Status of publication (accepted/peer-reviewed).**
> **Expected publication date.**

5. **Appendices (If Required by Guidelines)**
   - CV of author.
   - Acceptance letter from journal.
   - A copy of the manuscript (either in finished form or peer-reviewed version).

11.7 Crowdfunding (When There's a Will, There's a Way!)

Crowdfunding is a modern way of raising funds that has recently been gaining popularity.

- The most important part of crowdfunding is setting up an attractive campaign and sharing it with communities that might be interested.
- GoFundMe is one of the most recognized fundraising platforms; it is free to use, requiring only a small transaction fee charged per donation.
- Experiment is another platform but more concerned with raising funds for scientific research. You are required to pay a platform fee.
- (Make sure to check which platforms are available in your country)

References

1. Oxford learner's dictionaries. Definition of fund. Retrieved from: https://www.oxfordlearnersdictionaries.com/definition/english/fund_1?q=fund
2. Letpub. Journal submission and publication fees: A Review. Retrieved from: https://www.letpub.com/Journal_Submission_and_Publication_Fees_A_Review
3. Hopf H, Krief A, Mehta G, Matlin SA. Fake science and the knowledge crisis: ignorance can be fatal. R Soc Open Sci. 2019;6(5):190161.
4. AkJournals. Everything you need to know about article processing charges. Retrieved from: https://akjournals.com/page/214
5. IFIS. Open-access models. Retrieved from: https://ifis.libguides.com/journal-publishing-guide/open-access-models

6. Springer nature. Open-access agreement for Egypt. Retrieved from: https://www.springernature.com/gp/open-science/oa-agreements/egypt
7. Deal consortium. Agreements. Retrieved from: https://deal-konsortium.de/en/agreements
8. Elsevier. India. Retrieved from: https://www.elsevier.com/open-access/agreements/india
9. Wiley. JISC agreements. Retrieved from: https://authorservices.wiley.com/author-resources/Journal-Authors/open-access/affiliation-policies-payments/jisc-agreement.html
10. NHMRC. Open-access policy. Retrieved from: https://www.nhmrc.gov.au/about-us/resources/nhmrc-open-access-policy
11. ARC. Grant application. Retrieved from: https://www.arc.gov.au/funding-research/apply-funding/grant-application
12. DFG. Information sheet on publication assistance. Retrieved from: https://www.dfg.de/de/formulare-51-10-246726
13. JSPS. Open-access. Retrieved from: https://www.jsps.go.jp/j-grantsinaid/01_seido/08_openaccess/index.html#faq
14. QNL. List of publishers. Retrieved from: https://www.qnl.qa/en/open-access-at-qnl/open-access-agreements/list-of-publishers
15. Bill & Melinda Gates Foundation. Journal checker tool. Retrieved from: https://openaccess.gatesfoundation.org/how-to-comply/journal-checker-tool/

Peer Review: How to Reply to Reviewers

12

Mennatallah Alashker, Tasnim Awad,
Islam Mohammad Shehata, Omar Viswanath,
and Musa Aner

You are not always right, and they are not always right.

—Islam Mohammad Shehata

12.1 Introduction

For the manuscript to be accepted for publication in a journal, it has to go through a reviewing process known as a peer review. The USA invests $2 trillion US dollars globally to produce around three million peer-reviewed research articles a year [1]. This gives us a scale of the significance of ensuring scholarly research quality, legitimacy, and authenticity through reviewing. Even though there are some cases against peer review, it is still deemed the most reliable form of scientific evaluation.

M. Alashker (✉)
Radiology Resident at National Hepatology and Tropical Medicine Research Institute,
Cairo, Egypt

T. Awad
Intern doctor, Zagazig University Hospitals, Zagazig, Egypt

I. M. Shehata
Department of Anesthesiology, Faculty of Medicine, Ain Shams University Cairo,
Cairo, Egypt
e-mail: Islam.shehata@med.asu.edu.eg

O. Viswanath
Department of Anesthesiology, Creighton University School of Medicine, Phoenix, AZ, USA

Mountain View Headache and Spine Institute, Phoenix, AZ, USA

M. Aner
Section Chief, Pain Medicine, Center for Pain and Spine, Dartmouth Hitchcock
Medical Center, Lebanon, NH, USA

Anesthesiology Geisel School of Medicine, Dartmouth, NH, USA
e-mail: Musa.M.Aner@hitchcock.org

© The Author(s), under exclusive license to Springer Nature Switzerland AG 2025
I. M. Shehata, O. Viswanath (eds.), *How to Successfully Publish a Manuscript*,
https://doi.org/10.1007/978-3-031-92538-2_12

In this chapter, we will dissect the reviewing process and the specifics authors must know when it comes to their manuscript getting reviewed. We will also discuss how to effectively reply to reviewers' different comments to achieve constructive communication.

12.1.1 What Is Peer Review? (Fig. 12.1)

It is the process in which other researchers are given the role of reading and revising your work in which they assess the integrity of the paper in many different ways. Remember, in your journey as a researcher, your work will be subjected to peer review, and you will also have a chance to review other people's work. Peer review offers a community for the scientists in which established researchers review and validate your work, and in turn you could provide peer review for other researchers' work. It is a big part of developing an academic identity. A 2015 survey done by The Publishing Research Consortium shows that 82% of researchers asked were in agreement with the idea that there would not be any control over scientific communication if it were not for peer review [2].

12.1.2 Importance of Peer Review

Peer review is an essential system and an integral part of the research process. It is considered the last part of the research cycle. The process of reviewing serves as a quality control mechanism or a sieve that filters only scientifically sound research articles where the reviewers help the authors with their experience to authenticate the papers before publication. It ensures that most research articles being produced are:

- Authentic
- Coherent
- Credible
- Practical

12.1.3 Brief History of Peer Review (Fig. 12.2) [4, 5]

12.1.4 Who Is Responsible for Reviewing?

Those responsible for reviewing the manuscript are often experts in the field of the research topic (hence its name, peer review) with different academic ranks and professional backgrounds [6].

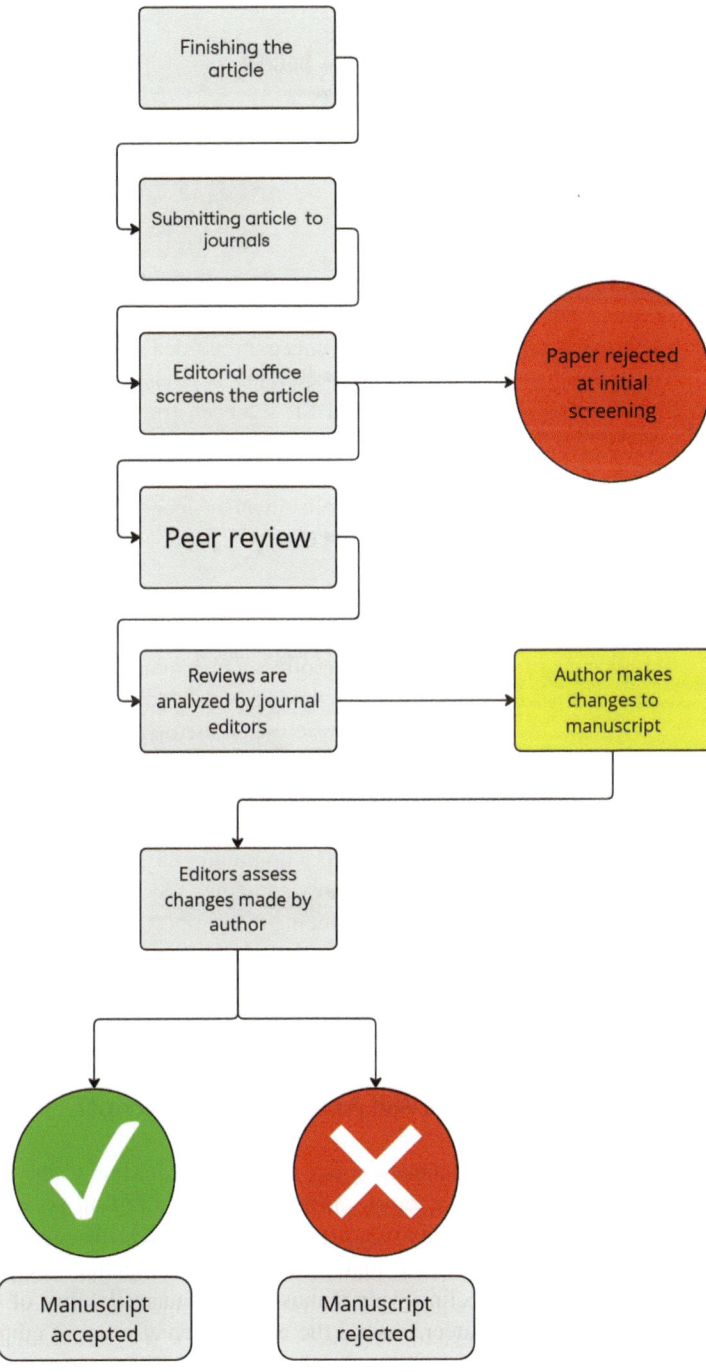

Fig. 12.1 Process of peer review [3]

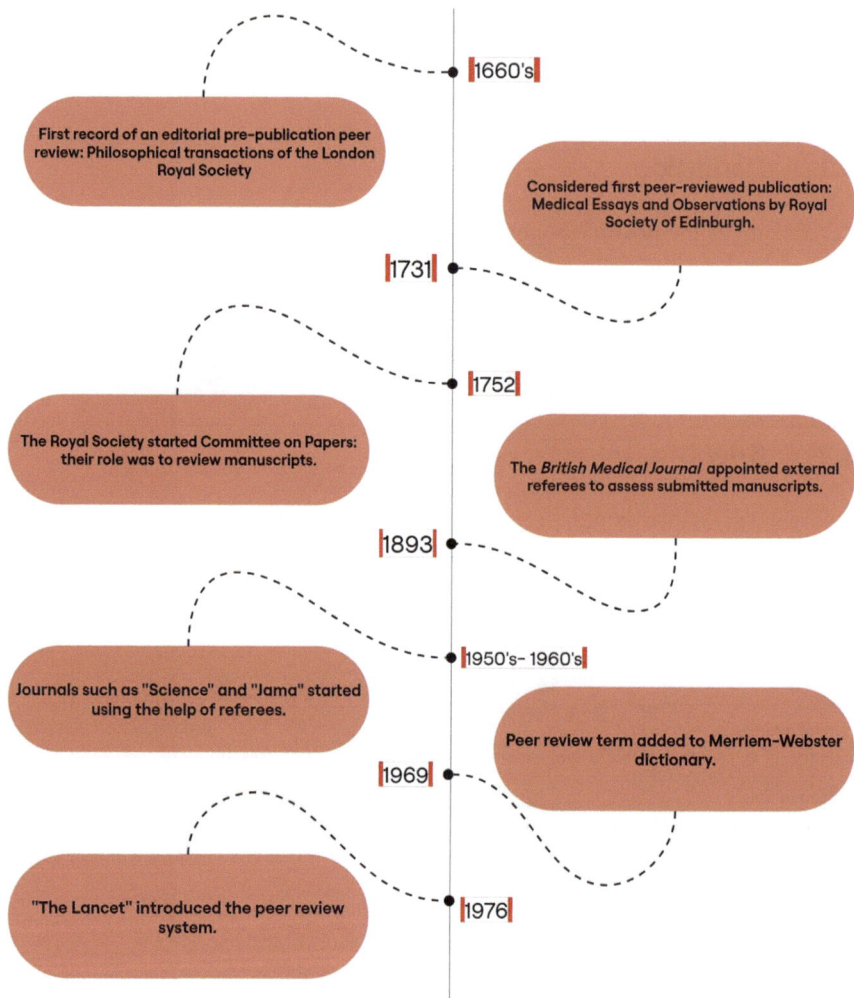

Fig. 12.2 History of peer review

12.1.5 How Many Reviewers Will Assess My Manuscript?

The number of reviewers to a manuscript is determined according to the journal's policies and guidelines. However, most journals assign two to four reviewers per manuscript. In case of unavailability of more than one reviewer, the editorial office asks the editor-in-chief to assess your manuscript or whether another reviewer can be recruited or whether to decline your manuscript for unavailability of enough reviewers. You can even volunteer to send the editor-in-chief a list of editors you and your fellow authors believe are fit to review your manuscript.

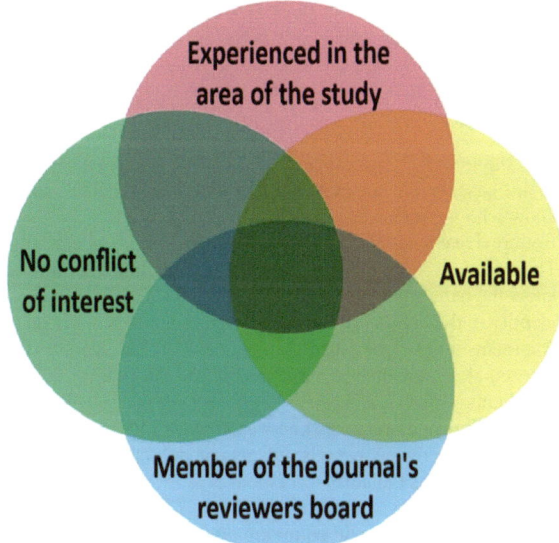

Fig. 12.3 Common criteria for choosing reviewers

12.1.6 What Are the Criteria for Choosing Reviewers? (Fig. 12.3)

The editor may even allow the authors to suggest reviewers who they think have the appropriate expertise in cases of open peer review. However, reviewers with a close connection or relationship to the listed authors, for example, working in the same institution, who have collaborated as past co-authors or family relatives, should not be recommended [7].

It is the ethical norm that the editor and the reviewers would not let the origins of a manuscript, the nationality, religious beliefs, gender, political views, other personal information about the authors, or any commercial considerations affect their decision to accept or reject the manuscript [8]. In some cases, to maintain transparency, the manuscript would be sent to the reviewer without author affiliation to maintain anonymity.

It is the ethical expectation that reviewers would decline the invitation to review if there is a conflict of interest. If they are unsure whether a relevant interest is involved, they inform the journal of any potential conflicts of interest and consult the journal [7].

12.1.7 What Are the Types of Peer Review?

The type of peer review used by a journal should be clearly stated either on:

- The invitation to review letter
- The journal's website

If there is still confusion about the peer-review process or a clarification on the journal's policy is needed, the journal's editors should be contacted with any inquiries.

Types of peer review [9]

Peer review	Description
Single-blind	In this case, reviewers know who the authors are, while authors would not know who the reviewers are
Double-blind	Both reviewers and authors do not know about the other entity
Open peer review	The reviewers know the names of the authors; however, the authors do not know the names of the reviewers unless the reviewers choose to reveal their identity to the authors. If the manuscript is published, the reviewer's comments will be published (anonymously unless they choose to sign their reports) alongside the main manuscript and the author's responses to the comments
Transparent review	Both authors and reviewers know each other's identities. If the manuscript is published, the review reports will also be published under the reviewer's names as well as the author's responses to reviewers' comments
Less common peer-review types	
Triple-blind	The author, reviewer, and editor are all unaware of each other's identities
Collaborative	More than one reviewer will work together to give one report about a paper
Post-publication review	In this type of review, unlike all the other types which are done before publication, this type is done after the paper has been published. This type does not take place alone but happens alongside any of the other types

12.2 How Do Reviewers Review?

12.2.1 How Do Reviewers Read Your Manuscript?

The reviewers' main job is to determine if the author's claims are supported and to help the journal decide whether the research is worthy of publication.

Reviewers will start by browsing your manuscript thoroughly and taking a look at all the figures. Sometimes, instead of reading the full manuscript, they will read the important sections and then read the rest of the full manuscript from beginning to end while taking notes, starting with reading **the abstract and the introduction** to get an idea of the manuscript, moving on to **the figures and tables** to see if they coincide with the **results, and** then finally reading **the conclusion**. Of course, throughout the whole process, they will be taking notes and then re-reading the manuscript multiple times for further evaluation.

A useful tip is to read your own manuscript as if you were a reviewer or send it to a reviewer you know personally before submission. This can give you an idea of how your manuscript will look to reviewers and might also make you reassess if there are any changes to be carried out.

12.2.2 How Is Your Manuscript Processed? [10]

After you submit your manuscript to the journal, it goes through several stages before it is reviewed.

1. Technical Check
   A general assessment is made by the secretary office where the paper is first read through to assess if it follows the journal guidelines or not. Common issues that arise are:
   - Incomplete submission
   - Failure to follow guidelines:
     – Word count
     – Artwork (resolution, format, etc.)
     – Filing (adding the author list in the manuscript)
   - Submissions to multiple journals simultaneously
   - Plagiarism
   - Lack of cover letter
   - Missing affiliation of contributing authors
   - Lack of images' permissions or credit
   - Ethical concerns
2. Assigning an Editor
   If the manuscript passed the technical check, the journal will then select an editor to become responsible for the manuscript. The editor must have an experience in the scope of the topic of the manuscript, have no conflict of interest, and be available to be assigned to the manuscript.
3. With Editor
   The editor chooses the reviewers and oversees the reviewing process. They determine the fate of your manuscript after assessing reviewers' feedback, and they have the right to reject it outright, which is known as a desk rejection.
4. Peer Review
   After passing through the previous stages which can take varying periods of time depending on the journal, the editor starts sending out invitations to reviewers they deem appropriate. The reviewers will be checking each individual part exclusively: the title, the abstract, the methodology, the discussion, the figures, the references, and comments for every section will be noted.

12.2.3 What Are the Reviewers' Criteria for Assessment? (Fig. 12.4)

These criteria will be in the form of questions that the reviewer will always have in mind and continually ask themselves throughout their first read-through of your manuscript.

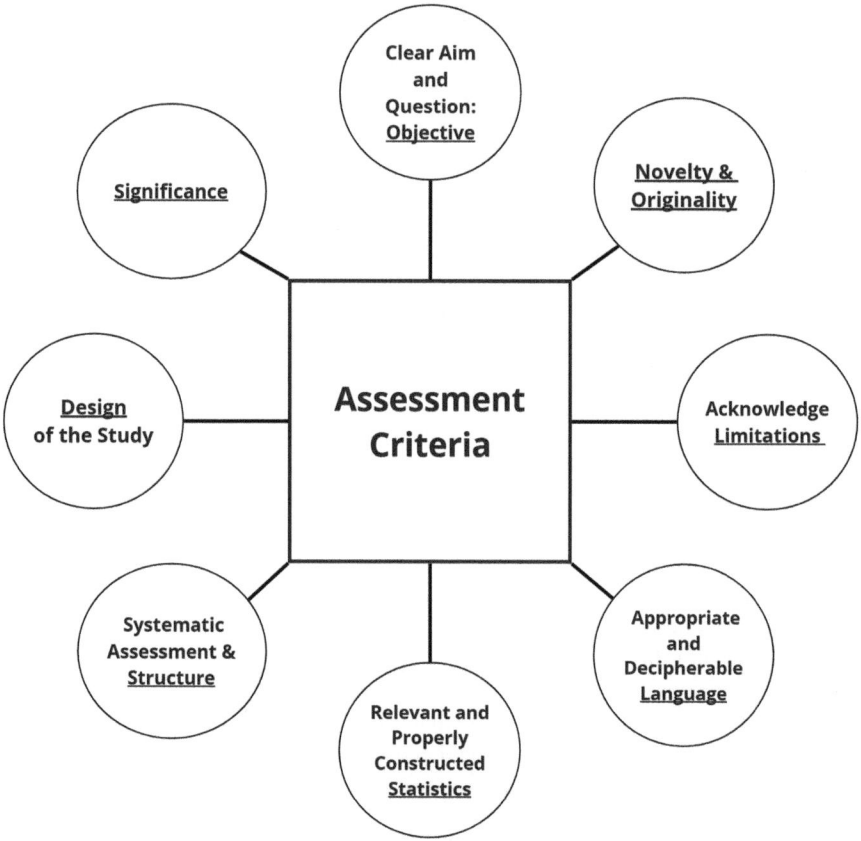

Fig. 12.4 Important reviewers' criteria

12.2.4 What Are the Assessment Forms Used by the Journals? (Fig. 12.5)

Most reviewers will have clear assessment forms employed by the journal to try standardizing the reviewing process (to make it equal and fair) among all different types and fields of research.

- Journals may employ a **structured questionnaire regarding** the manuscript for the reviewers to answer specific questions about the different sections of the research article.
- Some journals allow the reviewers to rate the different sections on a numerical scale.
- Most journals allow the reviewers to submit elaborate reports with their evaluations, feedback, and direct instructions to the authors on how to improve their work (revision).

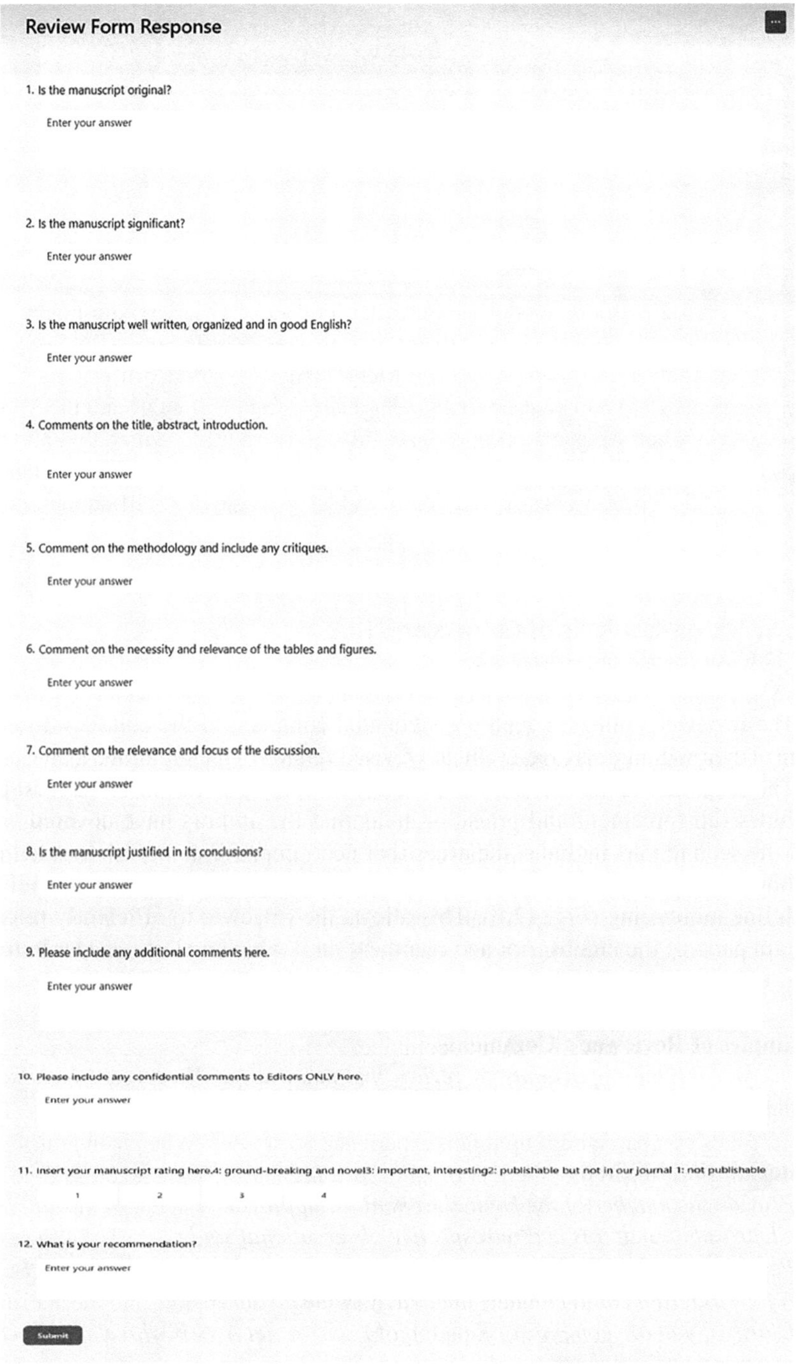

Fig. 12.5 An example of a review form

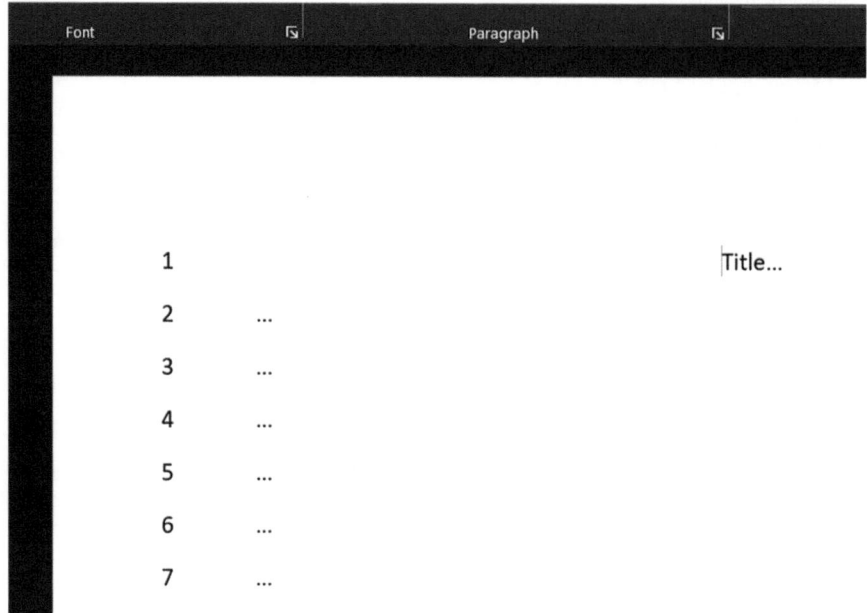

Fig. 12.6 An example of line numbering

The reviewers will also send a confidential comment to the editors informing them of their willingness to re-evaluate a revised version of the submitted manuscript.

The feedback of the reviewers will likely be split into two parts: the first part includes reinforcement and praise of areas that the authors have covered well, and the second part includes the areas that need improvement with instructions on how to do so separately. Some journals convert the manuscript into a pdf file with line numbering (Fig. 12.6). This allows the reviewer to efficiently refer to certain parts of the manuscript and comment on it which makes it easier to track their edits.

Examples of Reviewer's Comments
Thank you for the opportunity to review this manuscript. However, I have some comments to be shared with the authors regarding:

Materials and Methods
- Provide the number of the Ethical Committee approval.
- Clearly state that this is a retrospective observational study.
- Page 4, Line 52: "Each group included 50 patients." This sentence is misplaced.
- A flow diagram could enhance the clarity of the methods.
- Clarify if you are comparing X on top of Y and target A vs B with standard care.
- Define what "…" means.

Inclusion Criteria
- *Page 5, Line 8:* **Remove** *"clinically diagnosed with coronary heart disease" as it is redundant.*
- *Page 5, Line 15:* **Separate** *the exclusion criteria from the inclusion criteria.*
- *Page 5, Line 19:* **Define** *"significant cardiac surgery."*

Statistical Analysis
- *Include a sample size calculation or a power analysis.*
- *Justify the assumption of parametric distribution; otherwise, the use of a t-test can be misleading.*

Results
- *Page 8, line 23–37: These sentences are redundant. All data are reported in Table 1.*
- *Table 1: Report the median and interquartile range instead of the mean and standard deviation. Table 1 should include much more intraoperative and postoperative data such as ……….*
- *Page 9, Line 10: Provide the exact p-value.*

General Comments
The study is well-conceived and executed, providing valuable insights into the effectiveness of a ……. The findings are significant and relevant to both medical education and clinical practice.

The authors have done well in the following areas:

- *The **abstract** effectively summarizes the study, providing a clear overview of the background, methods, results, and conclusions.*
- *The **methodology** is described in detail, ensuring transparency and reproducibility.*
- *The **results** are presented clearly, with appropriate use of tables to summarize the data.*

However, the manuscript could benefit from a more detailed exploration of certain aspects, such as the sampling method, limitations, and specific recommendations for practice improvement. Addressing the identified areas for improvement would enhance the robustness and applicability of the findings.

Some of the areas for the improvement:

- *The manuscript lacks details on the pilot testing and validation of the survey instrument. Including this information would strengthen the study's reliability.*
- *The **sampling method** is not clearly described, which raises questions about the sample's representativeness.*
- *Including effect sizes and confidence intervals would enhance the interpretation of the findings.*

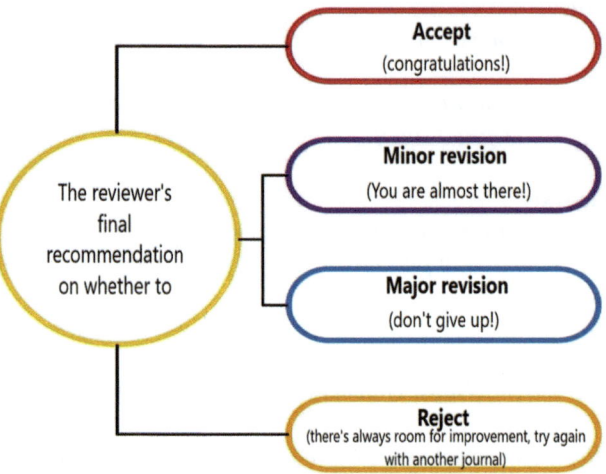

Fig. 12.7 Reviewers' recommendations on manuscripts

- *The discussion of limitations is brief. A more thorough examination of potential biases and generalizability issues is needed.*
- *The study would benefit from a deeper comparison with international literature to contextualize the findings.*
- *The authors should provide more specific recommendations for curriculum development based on their findings.*
- *The study's limitations are not fully discussed. Addressing issues such as self-report bias, potential non-response bias, and generalizability would strengthen the manuscript.*

12.2.5 What Are the Possible Outcomes? (Fig. 12.7)

12.3 How to Reply to Reviewers?

12.3.1 Always Remember the Following Before Replying to Reviewers

You are human and so is your reviewer.

You must remember that you are human and might make mistakes, but so is your reviewer!

Have trust in your text when you are reading your reviewer's comments. If you feel the need to defend your text or think there is more elaboration that needs to be made so your reviewer can understand the full image, please do not hesitate to do so. On the other far end of the spectrum, do not fall into the trap of thinking your reviewer is inexperienced or is unable to understand the text.

- **You should be:**
  - **Meticulous**: **Read the reviewer's comments very well.** If necessary, make a checklist with all the reviewer's comments for easier access and to make sure not to forget, skip, or overlook any of the comments [11].
  - **Organized:** Authors should be able to easily find everything in your manuscript without having to look everywhere for it. When signalling to a certain point you want the author to read, or a change you have made, make sure to include it word-for-word in your response to save the reviewer the hassle of looking through the text [11].
    Example:
    Reviewer comment: [Paste full comment here and italicize it.]
    Your response: "We agree with this comment, thank you for pointing this out. Therefore, we have… [Explain the change you have made and add the text you have revised. Then reference where in the edited manuscript this change is found. Mention the page number, the paragraph and the line].
- **Prompt**: Replying **promptly after reviewers send their comments** is essential to leave a good impression and to make the process quicker for both parties this would speed up the publication process.
- **Realistic**: **Know what to expect**. While not all the comments you will get are negative, most of the comments are probably bound to be. Do not take it too personally.
- **Smart**: Do not fall into the trap of thinking your reviewer is inexperienced or that they cannot understand your work. Be patient and polite when answering them, addressing all their points and their concerns [11].
- **Positive**: It is very important to approach the process of replying to reviewers with a positive attitude. Even though it might be daunting to have others scrutinize your work, think of their comments to your work as a gift that helps you improve your work.
- **Tactful:** Try to be tactful when explaining why you disagree with the reviewer. To support your argument and strengthen your point of view, you may use informative tables, statistics, and diagrams that you have not used in the original text [12].
- **Specific**: When you have been asked to make minor changes in the text, such as correcting a misspelled word, avoid replying with simple answers such as yes or no. Instead try to be more specific, for example, say: "We've corrected the typo." [11]
- **Concise**: While specificity is essential, try to also keep your responses concise and to the point.
- **Confident**: Reviewers are most likely very busy people, and for them to give your work a portion of their time is not something to be taken lightly. Other people's points of views can provide great insights and help to positively adjust the manuscript to reach its full potential.

- **Flexible**: Even if you do not agree with a small change suggested by a reviewer, it is usually better to just make the change if it is minor and will not significantly affect or redirect your work. It proves you are open to suggestions [12].

12.3.2 How to Write the Reply?

Always start by thanking the reviewer in the reply to every comment.

Before replying to the review, you must express your gratitude toward your reviewer for highlighting/ grabbing attention/ enhancing, etc., whether you are replying in agreement or disagreement, to a positive or a negative comment.

Thanking them is the first part of your action plan.

Examples:
In agreement:
1. *"We express thanks to reviewer for their constructive and insightful comments and suggestions. We believe that their feedback has significantly improved the quality of the manuscript."*
2. *"Thank you for giving me the opportunity to submit a revised draft of my manuscript."*
3. *"We appreciate the reviewer's careful review and constructive criticism."*
4. *"Thank you for bringing this point to our attention, we agree with this point and have updated the manuscript after carefully taking your suggestions into consideration."*
5. *"We thank the reviewer for these insightful observations; the problem has been fixed."*
6. In disagreement:
7. *"You have raised an important point here. However, we believe that …. would be more appropriate because…."*
8. *"Thank you for your suggestion. It would have been interesting to explore this aspect…, However, …."*
9. *"We thank you for your input on this point and while we appreciate the reviewer's feedback, we respectfully disagree for the following reasons…."*

Replying to the reviewer's edits **without neglecting** any part of it.

Make a **point-to-point** list of all the edits you have made (as mentioned above). When writing your point-to-point responses, include all of the reviewers' comments verbatim as they provided them. After each one of reviewer's comments, be sure to provide your response. To make things easier for you and for the editor, you can use a table [11].

Highlight all the changes in the body of the manuscript you have made, so they are easier to be found by the reviewers.

When disagreeing with the reviewer; make sure to **provide the evidence** needed to back up your claims, and remember to do so politely. Make sure to thoroughly explain your evidence.

When responding to one reviewer, make sure to respond as if they were the only reviewer you are replying to. Please refrain from mentioning other comments made by other reviewers.

Reread your revised manuscript to make sure you have integrated all the points that needed revision before sending it out.

Leave a good last impression by ending your letter with:
- A simple thank you for taking the time to review your manuscript.
- Showing that you are open to further suggestions and ready for changes (even if you are disagreeing with the reviewer).

12.3.3 Common Comments and How to Respond?

1. **Novelty and Originality of the Study**

    This is the first thing a journal must ensure before publishing. A research study must add to existing knowledge and explore the topic under a different light and contribute significantly and positively to the field of the study. The reviewer will likely compare your work to previously existing work, something the authors have done before starting the manuscript to find the gaps in knowledge they wanted to fill (literature review). However, the reviewer may be able to spot the repetition of information that authors may have overlooked due to their extensive experience in the field. The reviewer may also suggest improving the reference list to include relevant recent advances in the field of study as well as old historical developments.

    Example: *"The research presented in this manuscript lacks novelty and originality. The findings are largely incremental and do not significantly advance the field."*

    How to Respond?

    Response in Agreement: *"We appreciate the reviewer's insightful comment. Upon careful consideration of your point of view, we agree that the novelty of our research is limited. While our findings provide valuable insights into [specific aspect], they're not very diverse from the already existing literature.... [Insert the alterations you have made])".*

    Response in Disagreement: *"We would like to thank the reviewer for their constructive feedback on our manuscript. While we recognize that our research builds upon existing knowledge, we believe that it makes an important and significant contribution to the field by [Elaborate on the specific contribution done in the manuscript].*

    To further accentuate the novelty of our work, we have revised the manuscript to highlight the following points: **[List all the revision you have made and number them for quick access to the reviewer.]** *We believe that these revisions strengthen the overall impact of our work."*

2. **Title, Abstract, and Introduction**
   - **An example on title:** "Title is too long, it should be shortened."

     ***Response*:** "Thank you for your recommendation, we have rewritten the title to a more suitable length."
   - **An example on introduction**: "The topic is still vague, and the research question is not outlined properly in the introduction."

     ***Response*:** "Thank you very much for steering the attention towards this point, the introduction has been updated and the research question is made clear, such that [explain the changes made]."
   - **An example on abstract**: "The abstract seems a bit disjointed."

     ***Response*:** "We have taken this comment on board and have amended the introduction to the abstract, as this was also a comment pointed out by reviewer."
3. **Significance of the Study**

   Part of the reviewer's comments on your paper is to judge if it provides new knowledge, has practical and useful application, addresses a critical problem, and aims to solve it. The reviewer's background will be extensive enough to be able to evaluate the significance the study holds and comment accordingly.
   - **Example:**

   The reviewer could make this comment: *"I didn't find this study having an impact or a significant contribution to the literature or to science."*

   How to Respond?

   A *response* to a reviewer expressing how they do not think the manuscript is contributing to new knowledge would be:

   "Thank you for the time you have spent reviewing our manuscript. While we appreciate the reviewer's feedback, we have to respectfully disagree. We believe this study makes an important contribution to the field because [describe the questions answered by your study and its results or findings]."
4. **Methodology**

   Replicability and reproducibility of the study methods must be one of the authors' goals in any study to ensure the research's credibility and that the findings are not a result of chance, bias, or error. A research study is reproducible when the existing data is reanalysed using the same methodology (i.e., using the same analysis code, algorithms, and tools) and leads to the same results; this proves that the results of the study are reliable. Replicability refers to the conduction of the entire research process again under similar conditions, but not necessarily the same data, and still yielding the same results, it shows whether the study's conclusions are inclusive for different conditions, populations, or settings. By writing a transparent methodology section, the authors can achieve reproducibility when other researchers can achieve replicability [13].

   An example of a positive comment on the methodology would be *"the methodology is described in detail, ensuring transparency and reproducibility,"* to which the author would respond by thanking the reviewer.

   If the reviewer has suggested a different approach to your study's methods: **How to Respond?**

You can *respond* by saying: *"you have mentioned an important point here. However, after some deliberation, we have concluded that... [Methodology stated in the manuscript] would be more appropriate because.... [Offer your reasoning with clear supporting evidence.]"*

5. **Language, Grammar, and Punctuation**

   The language, grammar, and punctuation of a research paper must be up to the highest standards of clarity, professionalism, and precision. Being clear and precise in your language, particularly when describing methods, results, or conclusions will ensure the readers' understanding. Ambiguity can undermine the validity and reliability of your research. Typing mistakes must not be made to avoid confusion.

   – **Example 1:** *"P7 in line145, please pay attention to English writing style. Hiring a native speaker to check the integrity of the grammar and language of the paper."*

      Response: *"Thank you very much, we have carefully considered your suggestion and will be hiring a proof-reader to avoid any English mistakes."*

   – **Example 2:** *"(The findings suggested an increased risk of seizures during AD treatment, though the highest risk was before the initiations of ADs... Intriguingly...) remove the extra dot here."*

      Response: *"We would like to thank you for pointing this out. We have corrected the typo mentioned and removed the extra dot (highlighted on page 10, line 14)."*

   – **Example 3:** *"Some abbreviations without explanations are given in the abstract."*

      Response: *"We would like to thank the reviewer for highlighting this point. We have reviewed and reread the abstract and have defined any unknown abbreviations [insert changes made]"*

   – **Example 4:** *"Overall, the manuscript is well-written. However, it still has numerous grammatical errors that need to be fixed."*

      Response: *"Thank you for opening our eyes to this issue. After closely studying the manuscript, we have revised and corrected all the grammatical errors."*

6. **Statistics and Data**

   Example: *"The BMI data and its standard deviation are not clear to me. Does a BMI of 38 mean that all of these patients are obese? Or has this data been expressed with units of proportion not appropriate for the interpretation of these latter?"*

   Response: *"Thank you very much for your correction, there is indeed a problem with the BMI data, which is caused by a mistake in the calculation of BMI [BMI= Weight (kg) / Height (m)], and it has been modified [BMI= Weight (kg) / Height2 (m2)]."*

7. **Artwork**

   Example: *"Figure captions are generally insufficiently detailed throughout the manuscript."*

Response: *"Thank you for your nice reminder. We revised most of the figure captions to make them clearer and more detailed."*

8. **Limitations of the Study**

   To identify the limitations of a study could mean increased credibility of your work. Reviewers will be reading the Discussion section thoroughly to check for the research's limitations. The sample size, potential biases, time constraints, and limited resources are all possible limitations that the reviewer will seek out.

   Example: *"The authors haven't sufficiently addressed the limitations of the study. The results could have been changed by lack of diversity in the study population and an inadequate sample size. A more detailed argument of the limitations is required."*

   How to Respond?

   Responding in agreement: *"We thank the reviewer for highlighting the limitations of our study. We have added some changes to the manuscript to address the following limitations.... [Reference where in the edited manuscript this change is found. Mention the page number, the paragraph and the line.]"*

   Responding in disagreement: *"You have raised a crucial point here. Nonetheless, we believe that... [Provide your explanation with clear supporting evidence.]*

   Example: *"While the dataset is extensive, some of the data collection methods seem to have potential biases. The authors should address these concerns and discuss their potential impact on the findings."*

   Response: *"Thank you for your constructive comment. We have closely considered your remark on the data collection, and we have looked over the possible biases more closely and addressed them."*

9. **Results and Interpretation**

   Example: "The results given in Figure 3 and Table 5 appear to be contradictory. Please explain this contradiction and provide an explanation for the inconsistency."

   Response*:* *"thank you for pointing out this inconsistency. Upon further investigation, we realized that the discrepancy was due to a small error in the data analysis. We have corrected the error and updated Figure 3 and Table 5 with the edited results."*

10. **Conclusion**

    Example: A reviewer may comment: *"The conclusion is weak and lacks a clear statement of the study's overall significance. The authors should strengthen the conclusion by highlighting the key findings, discussing their implications, and suggesting future research directions."*

    Response: *"We would like to thank you for pointing this out to us, we agree with this comment. Therefore, we have reviewed and rewrote the conclusion. [Insert the newly written section]".*

11. **References**
    Example 1: *"Page 4- line 21 - referencing something that you are not sure whether or it is published somewhere else or not does not seem like it should be allowed"*
    Response: *"The text has now been updated to remove any ambiguity with other papers being developed."*
    Example 2*: "P, 6 L4: please provide reference for method used here"*
    Response: *"We would like to thank you for pointing this out, the reference has now been added."*
    Example 3: "**Some** *articles should also be incited, such as doi: 10.1007/s00464-023-10483-2, doi: 10.1186/s12871-023-02123-y"*
    Response: *"Thank you for directing our attention to this point***.*** We have added these two references in the revised manuscript." (Page 10, line 6)"*

12.4 How Can AI Help You in Replying?

Authors can benefit from AI when replying to reviewers' comments. Researchers should carefully review and edit AI-generated responses to ensure accuracy, clarity, and a personal touch.

1. To check plagiarism and get its percentage, AI can run an accurate plagiarism check.
2. If the reviewer has mentioned something you do not understand or is new to you in their comments, you can learn more about it by asking specific questions to AI.
3. AI can fix language problems and correct any grammar mistakes.
4. AI can suggest ways to make the writing more concise and easier to understand which also helps making the replying process faster and smoother (Fig. 12.8).
5. In case you want to paraphrase a response to make it more coherent or well structured, you can use the assistance of AI.

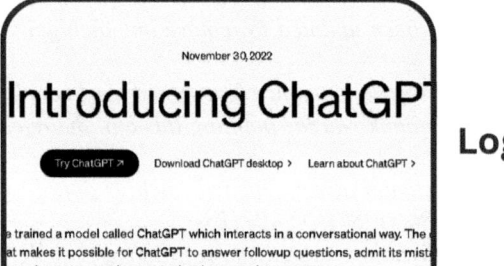
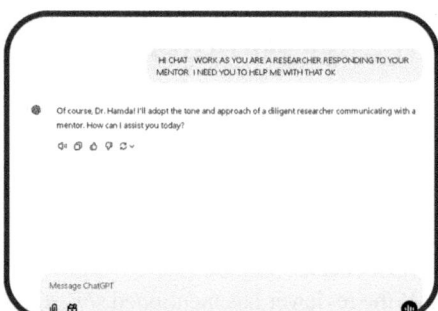

Fig. 12.8 An example of AI tool that can aid in replying

References

1. Tennant JP, Ross-Hellauer T. The limitations to our understanding of peer review. Research Integrity and Peer Review. 2020;5:1–14. https://doi.org/10.1186/s41073-020-00092-1.
2. Publishing Research Consortium Peer Review Survey 2015 Mark Ware Consulting. 2016.
3. BioMed Central. Peer review process. Biomedcentralcom. 2019. www.biomedcentral.com/getpublished/peer-review-process.
4. Spier R. The history of the peer-review process. Trends Biotechnol. 2002;20:357–8. https://doi.org/10.1016/S0167-7799(02)01985-6.
5. Al-Mousawi Y. A brief history of peer review. F1000. 2020. URL: https://blog.f1000.com/2020/01/31/a-brief-history-of-peer-review/
6. Tobias JD, Tumin D. The peer review process. Saudi J Anaesth. 2019;13:52. https://doi.org/10.4103/sja.SJA_544_18.
7. How to Peer Review | Publish your research | Springer Nature. URL: https://www.springernature.com/gp/authors/campaigns/how-to-peer-review
8. Hames I. COPE ethical guidelines for peer reviewers. 2013. URL: https://publicationethics.org/files/Ethical_guidelines_for_peer_reviewers_0.pdf
9. Wiley. Types of peer review | Wiley. Wileycom. 2019. URL: https://authorservices.wiley.com/Reviewers/journal-reviewers/what-is-peer-review/types-of-peer-review.html
10. (2024) Snapp - Submitting research - How to submit | Snapp homepage | Springer Nature. In: Springernature.com. https://www.springernature.com/gp/snapp/submitting/how-to-submit
11. Cushman M. How I respond to peer reviewer comments. Res Pract Thromb Haemost. 2023;7(2):100120. https://doi.org/10.1016/j.rpth.2023.100120. PMID: 37063759; PMCID: PMC10099302
12. N. C. How to respond to reviewers' comments: a practical guide for authors. 2020. Kaust.edu.sa. URL: https://anperc.kaust.edu.sa/docs/default-source/technical-writing/how-to-respond-to-reviewers-comments%2D%2D-a-practical-guide.docx?sfvrsn=f16dfea7_2
13. Nikolopoulou K. Reproducibility vs replicability | Difference & Examples. Scribbr. 2020. https://www.scribbr.com/methodology/reproducibility-repeatability-replicability/

How to Write A Case Report

Farah Mohamed Ismail, Ahmed Hashim, Islam Mohammad Shehata, Omar Viswanath, and Naum Shaparin

> *By the time there is a case study about your specialty, you are already too late*
>
> —Seth Godin

13.1 Definition

A comprehensive description concerning a specific patient's symptoms, signs, diagnosis, therapy, and follow-up that adds relevance to medical practice.

F. M. Ismail (✉)
Faculty of Medicine, Galala University, Suez, Egypt
e-mail: Farah.ismail@gu.edu.eg

A. Hashim
Cardiology Department, Faculty of Medicine, Ain Shams University, Cairo, Egypt

I. M. Shehata
Department of Anesthesiology, Faculty of Medicine, Ain Shams University Cairo, Cairo, Egypt
e-mail: Islam.shehata@med.asu.eg

O. Viswanath
Department of Anesthesiology, Creighton University School of Medicine, Phoenix, AZ, USA

Mountain View Headache and Spine Institute, Phoenix, AZ, USA

N. Shaparin
Montefiore Medical Center, Bronx, NY, USA
e-mail: NSHAPARI@montefoire.org

© The Author(s), under exclusive license to Springer Nature Switzerland AG 2025
I. M. Shehata, O. Viswanath (eds.), *How to Successfully Publish a Manuscript*,
https://doi.org/10.1007/978-3-031-92538-2_13

13.2 Importance

Case reports are frequently used as the initial line of evidence for new interventions or as warning signs that a problem exists with previously established ones. Case reports are important clinical "touch points" that can promote necessary clinical discourse, which can ultimately yield further studies.

13.3 Benefits

- Offers a novel contribution or presents findings that are seldom addressed in the existing medical literature.
- Case reports are typically the first type of manuscript a new researcher will successfully publish, and its format allows for learning the proper steps of the entire process.
- Can add to one's resume for applications to medical schools or trainings.
- Presented at conferences (poster, talk, etc.) as an abstract or an invited presentation.
- If truly novel, your case report might give rise to a new eponymous cure/disease/procedure like Pasteur's treatment, Alzheimer's disease, and Halsted's operation.
- As typically the first foray into clinical research, publishing a case report gives a sense of accomplishment, thus boosting confidence and motivation to continue to do more research.

Types (Fig. 13.1)

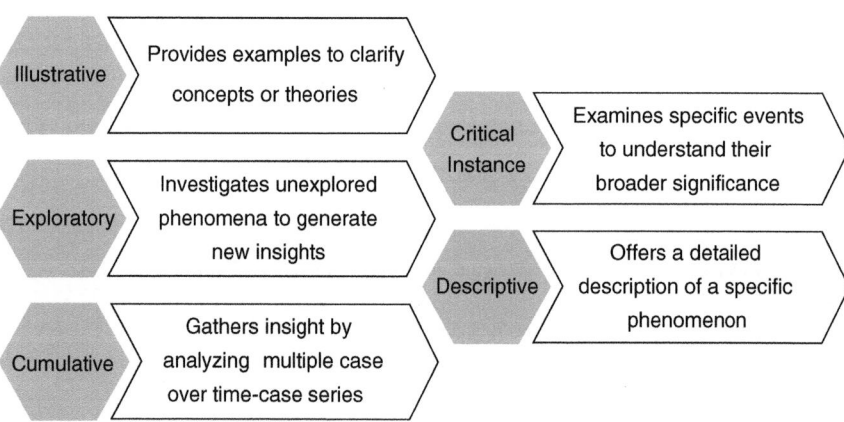

Fig. 13.1 An overview of the different types of case reports and their unique focus areas

> **Get to Know More:**
> Here are some examples about the different types of case reports:
> - **Illustrative**: https://journals.lww.com/aacr/abstract/2020/05000/pneumo-mediastinum,_pneumothorax,_pneumoperitoneum,.19.aspx
> - **Critical Instance:** https://pmc.ncbi.nlm.nih.gov/articles/PMC8314078/
> - **Descriptive**: https://www.jcvaonline.com/article/S1053-0770(21)00953-8/abstract

13.4 Characteristics of Publishable Case Reports (Fig. 13.2)

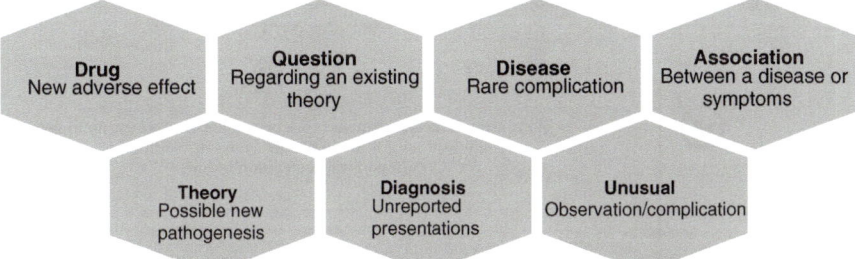

Fig. 13.2 Some characteristics of high-impact case reports

13.5 Preparing a Case Report

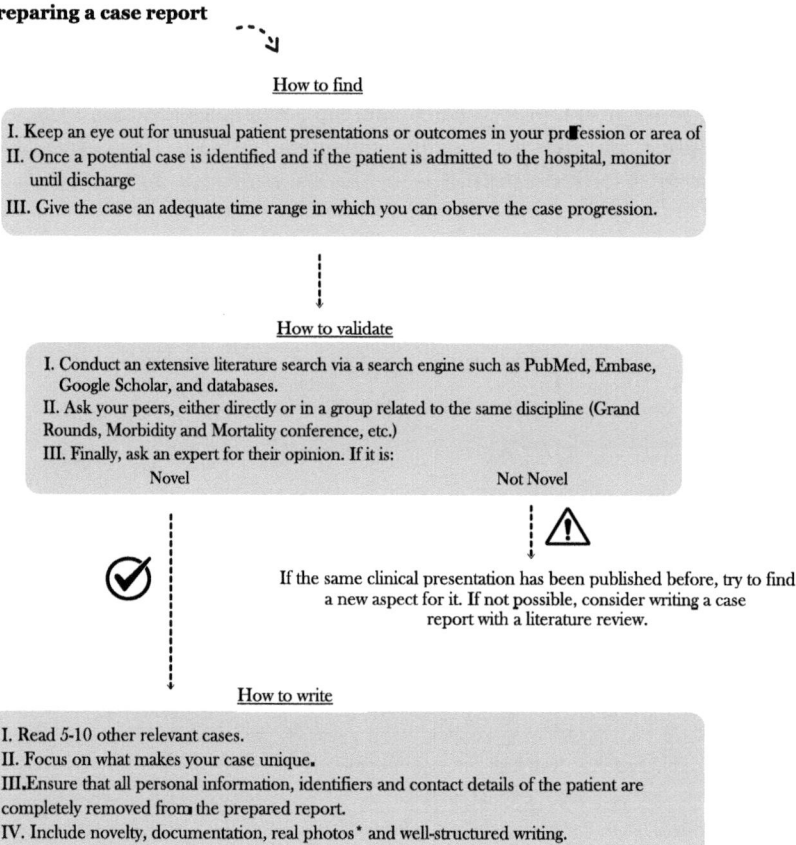

Prior to submission, ensure that the patient has provided informed consent for publication, as per institution policy, which should be explicitly documented in the report.

*Real photos increase the chances of publication acceptance, as in: https://journals.lww.com/aacr/abstract/2020/05000/pneumomediastinum,_pneumothorax,_pneumoperitoneum,.19.aspx (note that the face is covered up to preserve the patient's confidentiality).

13.6 Consent

13.6.1 Safe Harbor Method: De-identifying Patient Information

You must remove ALL identifying patient information, including:

1. Names.
2. All geographical subdivisions smaller than a State.
3. All elements of dates (except the year) for dates directly related to the patient.
4. All elements of dates (including the year) if the patient's age is above 89.
5. Telephone numbers and e-mail addresses.
6. Medical record numbers.
7. Biometric identifiers including finger and voice prints.
8. Full face photographic images and any comparable images.
9. Certificate/license numbers.

13.6.2 Respect Patients

Do not define patients by the disease process.

- Example: *diabetic/addict/epileptic/*etc.
- Instead: *Has diabetes/is addicted/suffers from epilepsy/*etc.

Do not blame the patients for an outcome.

- Example: *"The patient failed chemotherapy."*
- Instead: *"The patient's cancer didn't respond to chemotherapy."*

13.7 Structure of the Case Report

Make sure to follow the author's guidelines of your chosen journal.

Title
You should add "case report" / "case series" / "literature review" in your title. If you must, consider using a subtitle

Abstract
Should summarize the case, the issue it solves, and the message it delivers.

Keywords
Key terms including "case report" as one of them for indexing your work

Fig. 13.3 Visualization of the Funnel approach; from broad explorations to focused insights

What is known?

What is unknown?

Why was the study done?

Introduction
The funnel approach as shown in Fig. 13.3.

Case Presentation
A chronological description of:

- Chief complaint
- Patient description
- Medical history
- Physical examination results
- Results of pathological tests and other investigations
- Treatment plan
- Expected outcome of the treatment plan
- Actual outcome

Artwork

Include tables that present test results and figures of the clinical findings.

Discussion
Discuss the case presentation and the related literature and recognize any ambiguity or unexpected behavior that occurred.

Conclusion
A brief statement describing the message or lesson you want to express, as well as suggestions for future recommendations and clinical findings.

Acknowledgments
A place to recognize the mentors' contributors, the committee members, and the participants in the report.

You might not add it because case reports are usually short with low word maximums; if someone deserves to be acknowledged, they should be listed as an author.

13.8 ❌ The Don'ts of Writing

1. **Don't Omit Informed Consent**
   Do not forget to obtain and document patient consent, especially for identifying details like images or unique case histories.
2. **Don't Include Irrelevant Details**
   Avoid including too much unrelated information, which can dilute the focus of the case and make it harder to understand.
3. **Don't Forget to Highlight the Novelty**
   Avoid writing a generic report without clarifying why this case is unique or how it contributes to the medical literature.
4. **Don't Lack the Structure**
   Do not neglect a clear structure (e.g., introduction, case presentation, and discussion), as this will make the report disorganized and harder to follow.
5. **Don't Ignore Relevant Literature**
   Do not skip a thorough literature review, which helps frame the case within existing knowledge and highlights its significance.
6. **Don't Include Biased or Unsupported Claims**
   Avoid making conclusions or suggestions that are not backed by evidence or are based on limited experience.
7. **Don't Neglect Confidentiality**
   Do not include identifiable patient information (e.g., full name, exact dates) unless essential and permissible, as this breaches patient privacy.
8. **Don't Overlook Grammar and Spelling**
   Do not underestimate the importance of proofreading; errors in language can detract from credibility and clarity.
9. **Don't Skip the Discussion of Limitations**
   Do not avoid mentioning the limitations of the case report or any uncertainty, as it provides balance and transparency to your conclusions.

References

1. Carleton HA, Webb ML. The case report in context. Yale J Biol Med. 2012;85(1):93.
2. Moore K. A medical writing curriculum for internal medicine residents: using adult learning theory to teach formal medical writing and publication of case reports. MedEdPORTAL. 2015;1(11):10073.
3. Luciano G, Jobbins K, Rosenblum M. A curriculum to teach learners how to develop and present a case report. MedEdPORTAL. 2018;16(14):10692.
4. Alsaywid BS, Abdulhaq NM. Guideline on writing a case report. Urology Annals. 2019;11(2):126–131 [Internet].
5. Guidelines to writing a clinical case report. Heart Views [Internet] 2017;18(3):104. Available from: https://www.ncbi.nlm.nih.gov/pmc/articles/PMC5686928/.
6. Florek AG, Dellavalle RP. Case reports in medical education: a platform for training medical students, residents, and fellows in scientific writing and critical thinking. J Med Case Rep. 2016;10:1–3.

7. Shehata IM, Hashim RM. Pneumomediastinum, pneumothorax, pneumoperitoneum, and subcutaneous emphysema complicating extubation of a difficult airway using an airway exchange catheter: is oxygen insufflation innocent?: a case report. A&A Practice. 2020;14(7):e01228.
8. Sheata IM, Smith SR, Kamel H, Varrassi G, Imani F, Dayani A, Myrcik D, Urits I, Viswanath O, Taha SS. Pulmonary embolism and cardiac tamponade in critical care patients with COVID-19; telemedicine's role in developing countries: case reports and literature review. Anesthesiol Pain Med. 2021;11(2)
9. Bhandary SP, Shehata IM, Richter E, Klopman M. Lung isolation in the setting of vocal cord implantation. J Cardiothorac Vasc Anesth. 2022;36(8):3129–30.
10. Viswanath O, Simpao AF, Santhosh S. Atypical presentation of a pulmonary embolism in the perioperative setting. A A Case Rep. 2015;5(4):54–6. https://doi.org/10.1213/XAA.0000000000000174. PMID: 26275306
11. Viswanath O, Wilson J, Hasty F. Harlequin syndrome associated with multilevel intercostal nerve block. Anesthesiology. 2016;125(5):1045. https://doi.org/10.1097/ALN.0000000000001208.

Systematic Review: A Comprehensive Guide

14

Manar Ahmed Kamal, Batoul Mohamed Alaswad, Islam Mohammad Shehata, Omar Viswanath, and Sarang S. Koushik

14.1 Context

During the late 1970s and early 1980s, a collective of health service researchers in Oxford laid the foundation for evidence-based medicine by initiating a series of systematic reviews focused on the efficacy of healthcare interventions [1]. Systematic reviews have established themselves as a fundamental element of evidence-based practice, acting as an essential resource for consolidating research outcomes and facilitating informed decision-making across diverse disciplines [2]. By offering a methodical and transparent approach to the evaluation of existing literature, systematic reviews effectively connect raw research data with practical applications [3].

The concept of systematic review has evolved significantly since its inception. Originally developed within the medical field to address clinical questions, the

---

M. A. Kamal (✉)
Faculty of Medicine, Benha University, Benha, Egypt

B. M. Alaswad
Faculty of Medicine, Modern University for Technology and Information, Cairo, Egypt
e-mail: Batoul.99072@Medicine.mti.edu.eg

I. M. Shehata
Department of Anesthesiology, Faculty of Medicine, Ain Shams University Cairo, Cairo, Egypt
e-mail: islam.shehata@med.asu.edu.eg

O. Viswanath
Department of Anesthesiology, Creighton University School of Medicine, Phoenix, AZ, USA

Mountain View Headache and Spine Institute, Phoenix, AZ, USA

S. S. Koushik
Associate Professor of Anesthesiology, Valleywise Health Medical Center,
Creighton University School of Medicine, Phoenix, AZ, USA
e-mail: Sarang_koushik@dmgaz.org

© The Author(s), under exclusive license to Springer Nature Switzerland AG 2025
I. M. Shehata, O. Viswanath (eds.), *How to Successfully Publish a Manuscript*,
https://doi.org/10.1007/978-3-031-92538-2_14

methodology has expanded to include a wide range of disciplines, including social sciences, environmental studies, and education. Systematic reviews are particularly valuable in contexts where decisions need to be based on the best available evidence, such as policy-making, clinical guidelines, and program evaluations [4].

Systematic reviews are prevalent across various medical disciplines, with the highest frequency observed in neurology, followed by cardiology and surgery [5]. In these fields, systematic reviews occur at a rate ten times greater than in less prominent areas such as nursing or anesthesiology [5]. This disparity can be attributed to the established tradition of systematic reviews, which typically emphasize the effect size and prioritize data derived from experimental methodologies, such as randomized controlled trials (RCTs). Conversely, observational and interpretative research approaches are often either omitted from these reviews or regarded as lower-tier evidence [6].

14.2 Introduction

14.2.1 Definition

Systematic review serves as a recognized method for objectively cataloging pertinent evidence, evaluating its quality, and synthesizing findings [7].

14.2.2 Importance of Systematic Review

These reviews concentrate on elucidating what is understood and what remains uncertain regarding a particular question, often of significant policy importance [5]. Nonetheless, there is a lack of agreement on the scope of systematic reviews. Some may examine the effectiveness of treatments, diagnostic criteria, or epidemiological trends, while others might emphasize measurement techniques or the methodological robustness of primary research. The key characteristic of a systematic review is its provision of objective conclusions derived from the existing evidence, rather than a comprehensive account of all aspects related to the topic [5].

Unlike narrative reviews, which may be more subjective and less comprehensive, systematic reviews adhere to a predefined protocol that ensures a comprehensive and unbiased summary of the literature [5]. This approach involves a series of systematic steps designed to minimize bias and provide a clear and reliable synthesis of the evidence [5]. Moreover, systematic review increases the internal validity, reliability, and exploratory power of the findings. Finally, they support the precision of estimated values and enhance the statistical power to detect true associations.

14.3 Process of Conducting a Systematic Review Is Defined by Several Essential Components [8]

By following these principles, systematic reviews seek to deliver a comprehensive and clear synthesis of evidence that can guide both practice and policy [8] (Fig. 14.1).

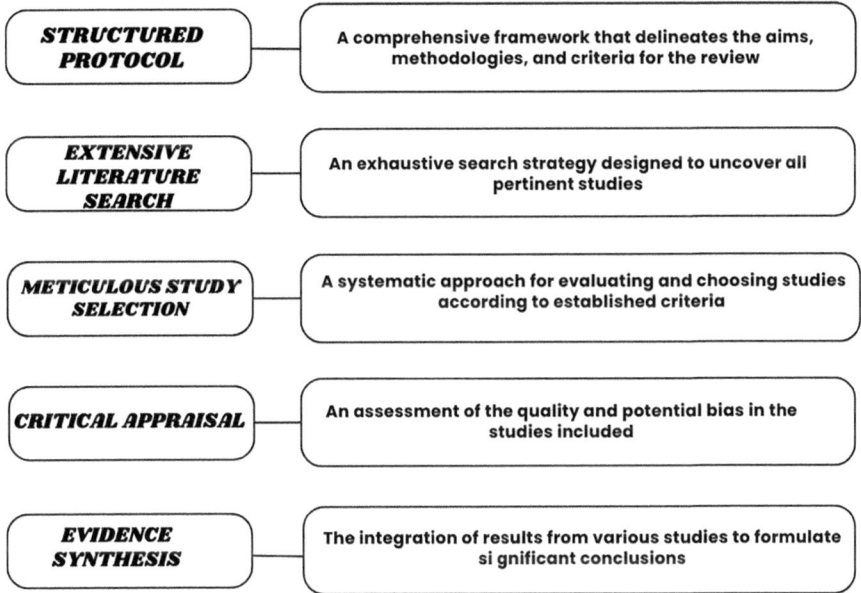

Fig. 14.1 Systematic essential components

14.4 Steps of Systematic Review

14.4.1 Formulating the Research Question

The initial phase of conducting a systematic review involves the development of a precise and well-defined research question. This question serves as the foundation for the entire review process and must be specific, pertinent, and achievable. The PICO framework—comprising Population, Intervention, Comparison, and Outcome—is frequently employed to organize clinical inquiries, although alternative frameworks may be applicable for various research question types. [9]

Example For a systematic review on the neurological side effects of an anesthetic drug (ketamine), the proposed research question was:
   Ketamine: Pro or antiepileptic agent? A systematic review [10].

14.4.2 Developing a Review Protocol

After formulating the research question, the subsequent phase involves creating a comprehensive review protocol. This protocol delineates the aims of the review, specifies the criteria for inclusion and exclusion, details the search strategy, describes the methods for data extraction, and outlines the intended analytical techniques.

[11] The protocol functions as a guiding framework for the review process, promoting both consistency and transparency. You can use the following links which are tailored specifically for systematic reviews:

- **PROSPERO: https://www.researchregistry.com/**
- **INPLASY: https://inplasy.com/**
- **Research Registry: https://www.crd.york.ac.uk/prospero/**

However, there are more versatile websites, accommodating a wide range of study types [12]:

- Protocols.io https://www.protocols.io/blog/peer-reviewed-protocols-a-new-initiative-between-protocolsio
- OSF Registries https://www.cos.io/products/osf-registries

The registration forms for systematic review registries typically include 24 to 28 mandatory fields. In contrast, OSF Registries offer users the choice between an "Open-ended-registration" option or various predefined forms, with the former enabling authors to present a narrative overview of their review [12]. All registries facilitate the search for registered systematic reviews; however, the effectiveness and functionality of these tools differ. PROSPERO generally offers the most comprehensive search capabilities, although recent evaluations have indicated that its search tools may not be optimal [12] (Fig. 14.2).

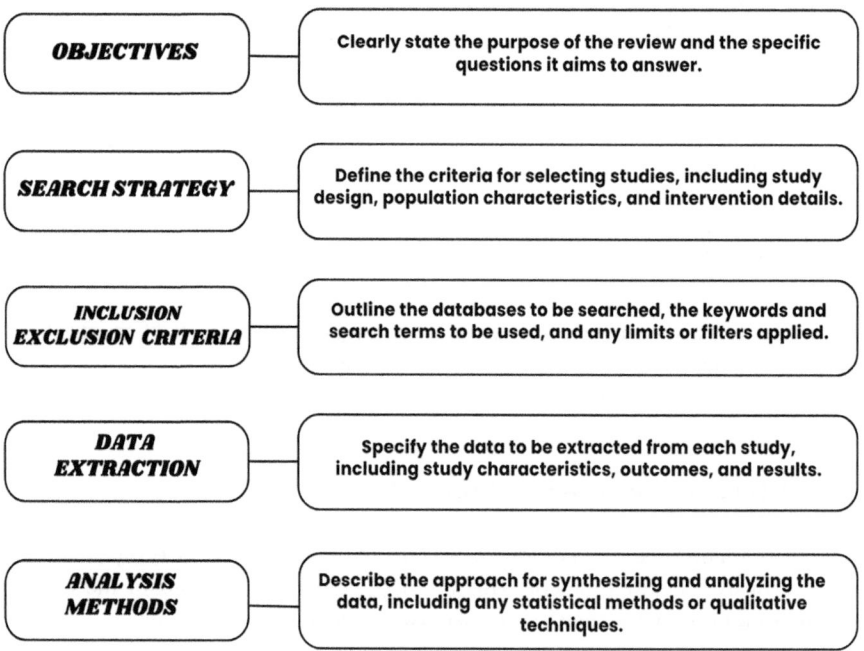

Fig. 14.2 Components of review protocol

14 Systematic Review: A Comprehensive Guide

Key Components of Review Protocol

14.4.3 Conducting Comprehensive Literature Search

An extensive literature review is essential for uncovering all pertinent studies associated with the research inquiry. This process entails exploring various databases, including PubMed, Cochrane Library, and Scopus, by employing a mix of keywords, subject terms, and Boolean operators. [13] Furthermore, manually reviewing reference lists and reaching out to specialists in the domain can aid in discovering additional relevant research.

Steps in the Search Process

14.4.4 Study Selection and Screening

After conducting the search, the next step is to screen and select studies for inclusion in the systematic review. This process involves two stages: initial screening and full-text review (Fig. 14.3).

14.4.4.1 Initial Screening Steps
A). **Title and Abstract Screening:** Assess the titles and abstracts of the retrieved studies to determine their relevance to the research question.

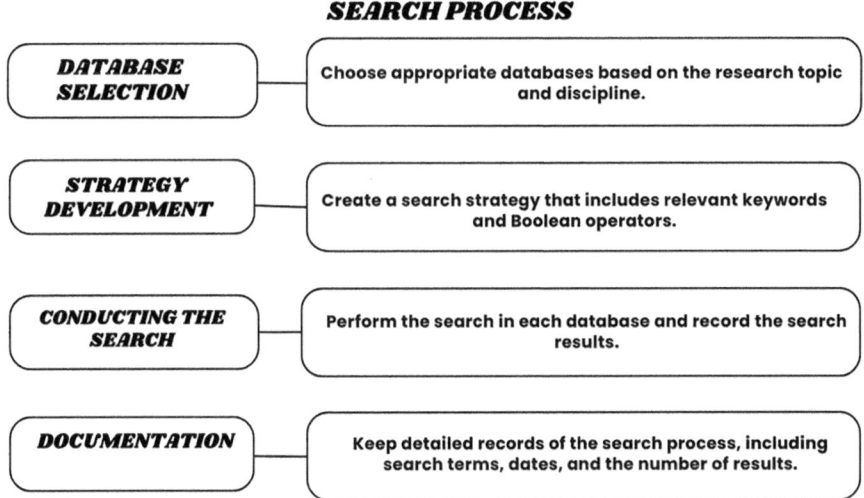

Fig. 14.3 Steps of the search process

14.4.4.2 Key Steps in Title and Abstract Screening [14]

1. **Establish Inclusion/Exclusion Criteria:** Clearly outline the specific types of studies that will be considered for inclusion, such as population characteristics, interventions, comparisons, and outcomes.
2. **Assess Titles:** Conduct a rapid evaluation of each title to determine its relevance. Titles that do not align with the established criteria may be excluded without further examination.
3. **Analyze Abstracts:** For titles deemed relevant, review the abstract to evaluate the study's objectives, methodologies, and results in relation to your criteria.
4. **Record Decisions:** Maintain a comprehensive log of articles that are included or excluded, along with justifications for exclusions, to ensure transparency in the process.

B). **Full-Text Screening:**

- **Eligibility Assessment:** Review the full text of the selected studies to confirm their eligibility based on the inclusion criteria.
- **Consensus:** Ideally, **two or more** independent reviewers perform the screening to minimize bias and resolve any discrepancies.

14.4.4.3 Key Steps in Full-Text Screening [11]

1. **Acquire Full Texts:** Secure the complete articles for all studies that have successfully passed the preliminary screening phase.
2. **Evaluate Against Criteria:** Conduct a thorough examination of each article to assess its methodology, findings, and relevance to your research inquiry.
3. **Record Inclusion/Exclusion:** Maintain comprehensive records of which articles are accepted, and which are rejected, along with the rationale for each choice.
4. **Data Collection**: For the studies that are included, extract pertinent data for analysis, which may encompass sample sizes, outcomes, and effect sizes.

14.4.4.4 Supporting Tools [2, 3, 13]

Rayyan: An online platform that enables researchers to collaborate effectively on article screening, allowing for tagging and organization of studies.

Covidence: A systematic review software designed to enhance the screening workflow, aiding in the management of inclusion/exclusion criteria and data extraction.

EndNote: Although primarily a reference management application, it can also support the organization and screening of literature.

Zotero: A complimentary tool that assists in gathering and organizing research articles, featuring capabilities for tagging and categorization.

Mendeley: Like EndNote, Mendeley assists in organizing references and offers features for annotating and sharing full-text articles.

Excel or Google Sheets: Basic spreadsheet applications that can serve as effective tools for tracking titles, abstracts, and screening decisions.

14.5 Data Extraction

Data extraction involves systematically collecting relevant information from each included study. This process is critical for ensuring that the data is accurate and comparable across studies. A standardized data extraction form or software tool can help organize and manage the data.

14.5.1 Data to Extract

- **Study Characteristics:** Author, publication year, study design, sample size, and setting.
- **Intervention Details:** Description of the intervention, dosage, duration, and comparison.
- **Outcomes:** Primary and secondary outcomes, measurement methods, and results.

14.5.2 Assistance Tools [15]

Data Extraction Templates: Numerous researchers create personalized templates tailored to their unique requirements. These templates can be formatted in various applications (such as Excel or Word) and must encompass all essential fields for thorough data collection.

RevMan (Review Manager): Created by Cochrane, RevMan is intended for the preparation and management of systematic reviews. It features tools for data entry and analysis, specifically designed for Cochrane reviews. https://login.cochrane.org/

Covidence: Covidence is a platform specifically crafted for systematic reviews, providing tools for data extraction that facilitate the process, enabling users to efficiently input and organize data from multiple studies. https://www.covidence.org/

1. **AI tools** (Table 14.1)

Table 14.1 AI tools help in systematic review conduction

Tool	Link	Character
Elicit	https://elicit.com/	Automate time-consuming research tasks like summarizing papers, extracting data, and synthesizing your findings
Scispace	https://scispace.com/	An incredible (AI-powered) tool to help you understand research papers better. It can explain and elaborate most academic texts in simple words
Scite	AI for Research | Scite	AI assistant or search the literature to transform the way you discover, evaluate, and understand research on any topic
Notebook LM	https://notebooklm.google/	The ultimate tool for understanding the information that matters most to you
ChatPDF	ChatPDF – Chat with any PDF – #1 Free PDF AI	Summarize and answer questions for free
Litmaps	https://app.litmaps.com/	Litmaps changes the way researchers search for papers and conduct literature reviews. Using the citation network and advanced search features, Litmaps help you find the most important papers on your topic faster
Connected papers ai	Connected Papers | Find and explore academic papers	With connected papers, you can just search and visually discover important recent papers. No need to keep lists
Researchrabbit	https://www.researchrabbit.ai/	It is intuitive and tremendously useful. That saves a lot of time identifying the most influential papers in a topic/field

14.6 Assessing Study Quality and Risk of Bias

Assessing the quality and potential bias in the studies incorporated into the review is crucial for understanding the trustworthiness of its conclusions. A range of instruments and checklists exists for evaluating study quality, including the Cochrane Risk of Bias Tool designed for randomized controlled trials [16] and the Newcastle-Ottawa Scale tailored for observational studies. [17]

14.6.1 Main Assessment Areas

- **Selection Bias:** Evaluate how participants were selected and whether the selection process was unbiased.
- **Performance Bias:** Assess whether the intervention was administered consistently across studies.
- **Detection Bias:** Determine if outcome assessment was conducted in a blind manner.
- **Attrition Bias:** Review how missing data was handled and whether there was differential loss to follow-up.

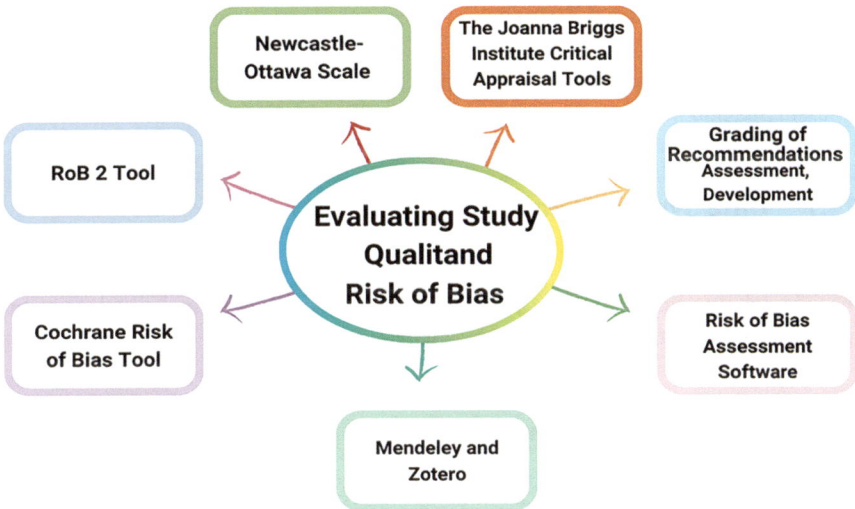

Fig. 14.4 Tools for evaluating study quality and risk of bias

14.6.2 Tools for Evaluating Study Quality and Risk of Bias [16–18] (Fig. 14.4)

1. **Cochrane Risk of Bias Tool**: This is a widely recognized instrument that establishes a framework for evaluating the risk of bias in randomized controlled trials (RCTs). It examines various domains, including selection bias, performance bias, detection bias, and reporting bias.
   LINK: https://methods.cochrane.org/bias/risk-bias-tool
2. **RoB 2 Tool:** An enhanced iteration of the original Cochrane Risk of Bias Tool; this version is specifically tailored for RCTs. It provides a systematic method for assessing risk across multiple domains.
3. **Newcastle-Ottawa Scale (NOS):** This tool is utilized for evaluating the quality of nonrandomized studies, particularly cohort and case–control studies. It emphasizes selection, comparability, and outcome assessment.
4. **The Joanna Briggs Institute (JBI) Critical Appraisal Tools**: This collection of tools is designed for assessing the quality of various study designs, encompassing qualitative research, randomized trials, and observational studies.
5. **GRADE (Grading of Recommendations, Assessment, Development, and Evaluations):** This system is employed to rate the quality of evidence and the strength of recommendations. It aids in determining the level of confidence to be placed in the findings of studies included in a review.
6. **Risk of Bias Assessment Software:** Software such as RevMan (Review Manager) incorporates features for documenting risk of bias assessments as an integral part of the review process.

14.6.3 Mendeley and Zotero

These reference management tools can also serve to store and organize quality assessments and notes pertinent to risk of bias (Fig. 14.4).

14.7 Data Synthesis and Analysis

Data synthesis involves combining the findings from the included studies to address the research question. This can be done through quantitative methods, such as meta-analysis, or qualitative methods, such as narrative synthesis.

14.7.1 Quantitative Synthesis (Fig. 14.5)

A. **Effect Size Calculation:** Calculate summary statistics, such as mean differences or odds ratios, to determine the overall effect of the intervention.
B. **Heterogeneity Assessment:** Evaluate the variability between studies using statistics like I^2 to assess how much the results differ from each other (check the chapter of meta-analysis).
C. **Sensitivity Analysis:** Test the robustness of the results by examining different assumptions or subsets of data.

14.7.2 Qualitative Synthesis

A. **Narrative Summary:** Provide a descriptive summary of the findings from the included studies, highlighting patterns and trends.
B. **Thematic Analysis:** Identify and summarize key themes and insights from the qualitative data.

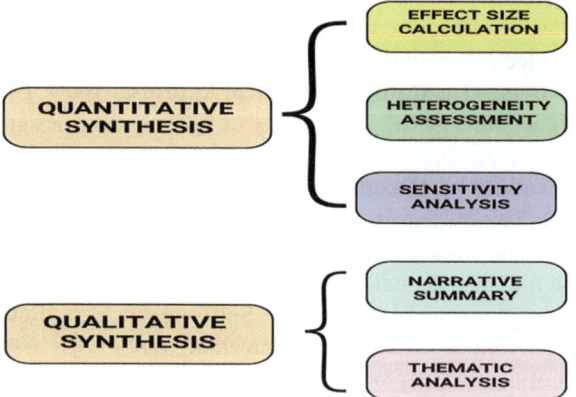

Fig. 14.5 Data synthesis and analysis methods

14.7.3 Meta-analysis (Please Check the Meta-analysis Chapter [19, 20] References Reorder!!!!)

Definition: Meta-analysis is a statistical technique that combines the results of multiple studies to derive a pooled estimate of effect size or to evaluate the consistency of findings across studies. It is often used in systematic reviews to summarize evidence, increase statistical power, and identify patterns or trends that may not be apparent in individual studies.

14.7.4 Key Steps in Conducting a Meta-analysis [19]

1. **Define the Research Question**: Clearly outline the question and objectives of the meta-analysis.
2. **Conduct a Systematic Review**: Gather and screen relevant studies using established methods (e.g., PRISMA guidelines).
3. **Data Extraction**: Collect relevant data from included studies, such as effect sizes, sample sizes, and study characteristics.
4. **Assess Study Quality**: Evaluate the quality and risk of bias of the included studies.
5. **Statistical Analysis**: Use statistical methods to combine effect sizes, which may involve fixed-effect or random-effects models depending on the heterogeneity of studies.
6. **Interpret Results**: Analyze and discuss the pooled results, addressing implications, limitations, and potential biases.

14.7.5 Helping Tools for Meta-analysis [19]

1. **RevMan (Review Manager)**:
   - Developed by Cochrane, this software is designed for preparing and maintaining systematic reviews and includes functions for performing meta-analysis.
2. **Comprehensive Meta-Analysis (CMA)**:
   - A widely used software specifically for meta-analysis that allows users to conduct various statistical analyses and visualize results.
3. **Stata**:
   - A powerful statistical software that includes meta-analysis commands for a variety of effect size measures and allows for complex data manipulation.
4. **R and R Packages (e.g., meta, metafor)**:
   - R is a free statistical computing environment, and packages like meta and metafor are tailored for conducting meta-analyses with extensive flexibility.
5. **OpenMeta[Analyst]**:
   - A free and open-source software for meta-analysis that allows users to analyze data and create forest plots.

6. **Cochrane Library**:
   - Offers access to systematic reviews and meta-analyses, providing a repository of high-quality evidence that can inform your analyses.
7. **Excel or Google Sheets**:
   - While not specialized software, these tools can be used to manage data extraction and perform basic calculations for effect sizes.

14.8 Reporting and Dissemination

The final step is to report the findings of the systematic review. The report should be comprehensive, transparent, and adhere to established guidelines such as PRISMA (Preferred Reporting Items for Systematic Reviews and Meta-Analyses) [8, 11].

14.8.1 Report Components

1. **Abstract:** A concise summary of the review's objectives, methods, results, and conclusions.
2. **Introduction:** Background information, research question, and objectives of the review.
3. **Methods:** Detailed description of the review process, including search strategy, study selection, data extraction, and analysis methods.
4. **Results:** Summary of the findings, including tables, figures, and statistical results.
5. **Discussion:** Interpretation of the results, implications for practice, limitations of the review, and recommendations for future research.

14.8.2 Dissemination

A. **Publication:** Submit the review to a peer-reviewed journal for publication.
B. **Presentation:** Present findings at conferences or seminars to share insights with the research community.
C. **Policy Briefs:** Prepare summaries for policymakers or practitioners to facilitate the application of findings in real-world settings.

Use the following AI tools for presenting your paper:

- **Beautiful.ai**
  AI-powered templates, automatic formatting, and design suggestions.
- **Tome.ai**
  AI-generated slides based on text prompts; integrate multimedia like videos and images.
- **Canva**
  Customizable templates, drag-and-drop interface, AI design suggestions, and animations.

14.9 Conclusions

Systematic reviews are indispensable tools for synthesizing research evidence and guiding evidence-based practice. By following a rigorous and transparent process, researchers can ensure that their reviews provide reliable and comprehensive summaries of the available evidence. This structured approach not only enhances the validity of the findings but also contributes to informed decision-making and the advancement of knowledge across various fields.

References

1. Greenhalgh T. Effectiveness and Efficiency: Random Reflections on Health Services. Vol. 328, BMJ : Br Med J 2004. p. 529.
2. Higgins JPT, Thomas J, Chandler J, Cumpston M, Li T, Page MJ, et al. Cochrane handbook for systematic reviews of interventions. Wiley; 2019.
3. Page MJ, McKenzie JE, Bossuyt PM, Boutron I, Hoffmann TC, Mulrow CD, et al. The PRISMA 2020 statement: An updated guideline for reporting systematic reviews. PLOS Med [Internet]. 2021 18(3):e1003583. Available from: https://dx.plos.org/10.1371/journal.pmed.1003583.
4. Higgins JP. Cochrane handbook for systematic reviews of interventions. Cochrane Collab John Wiley Sons Ltd.; 2008.
5. Hansen H, Trifkovic N. Systematic reviews—questions. Methods and Usage. 2013;1:62.
6. Evans D, Pearson A. Systematic reviews: gatekeepers of nursing knowledge. J Clin Nurs [Internet]. 2001;10(5):593–9. Available from: https://onlinelibrary.wiley.com/doi/10.1046/j.1365-2702.2001.00517.x.
7. https://www.google.com/search?q=systematic+review+definition&client=safari&sca_esv=427fd9fc564ca397&channel=iphone_bm&sxsrf=ADLYWIIiQuch7U0914rtTihqypBDD1g5Sw%3A1733261585233&source=hp&ei=EX1PZ9utC-ydigPopXvmA4&iflsig=AL9hbdgAAAAAZ0-HIUrVdJnezDt4X951qAbsIZSyPrs-&ved=0ahUKEwib-tnIxoyKAxXszgIHHaLKG-MQ4dUDCBc&uact=5&oq=systematic+review+definition&gs_lp=Egdnd3Mtd2l6IhxzeXN0ZW1hdGljIHJldmlldyBkZWZpbml0-aW9uMgUQABiABDIFEAAYgAQyBRAAGIAEMgYQABgWGB4yBhAAGBYYHjIGEAAYFhgeMgYQABgWGB4yBhAAGBYYHjIGEAAYFhgeMgYQABgWGB5Ig-i1Q0wNY6SpwAXgAkAEAmAGJAqAB4CKqAQYwLjI3LjG4AQPIAQD4AQGYA-hygAtoiqAIKwgIHECMYJxjqAsICChAjGIAEGCcYigXCAgsQABiABBixAxiDAcI-CERAuGIAEGLEDGNEDGIMBGMcBwgILEAAYgAQYsQMYigXCAg4QLhiAB-BixAxjRAxjHAcICDhAuGIAEGLEDGIMBGIoFwgIIEC4YgAQYsQPCAgsQLhiAB-BixAxiDAcICERAuGIAEGLEDGNEDGMcBGIoFwgIIEAAYgAQYsQPCAg4QABi-ABBixAxiDARiKBcICBRAuGIAEwgIIEAAYgAQYywGYAwySBwYxLjI2LjGgB7W_AQ&sclient=gws-wiz
8. Page MJ, McKenzie JE, Bossuyt PM, Boutron I, Hoffmann TC, Mulrow CD, et al. The PRISMA 2020 statement: an updated guideline for reporting systematic reviews. PLoS Med. 2021;18(3)
9. Higgins JPT. Cochrane handbook for systematic reviews of interventions version 5.0. 1. The Cochrane Collaboration. http//www cochrane-handbook org. 2008.
10. https://scholar.google.com/citations?view_op=view_citation&hl=en&user=tIbA7HQAAAAJ&citation_for_view=tIbA7HQAAAAJ:M3ejUd6NZC8C
11. Moher D, Liberati A, Tetzlaff J, Altman DG. Preferred reporting items for systematic reviews and meta-analyses: the PRISMA statement. PLoS Med [Internet]. 2009;6(7):e1000097. Available from: https://dx.plos.org/10.1371/journal.pmed.1000097.

12. Solla F, Bertoncelli CM, Rampal V. Does the PROSPERO registration prevent double review on the same topic? BMJ Evidence-Based Med [Internet]. 2021;26(3):140. Available from: https://ebm.bmj.com/lookup/doi/10.1136/bmjebm-2020-111361.
13. Boutron I, Page MJ, Higgins JP, Altman DG, Lundh A, Hróbjartsson A. Considering bias and conflicts of interest among the included studies. In: ochrane Handbook for Systematic Reviews of Interventions [Internet]. Wiley; 2019. p. 177–204. Available from: https://onlinelibrary.wiley.com/doi/10.1002/9781119536604.ch7.
14. Ng L, Pitt V, Huckvale K, Clavisi O, Turner T, Gruen R, et al. Title and abstract screening and evaluation in systematic reviews (TASER): a pilot randomised controlled trial of title and abstract screening by medical students. Syst Rev [Internet]. 2014;3(1):121. Available from: https://systematicreviewsjournal.biomedcentral.com/articles/10.1186/2046-4053-3-121.
15. Munn Z, Tufanaru C, Aromataris E. SYSTEMATIC REVIEWS, Step by Step Data Extraction and Synthesis The steps following study selection in a systematic review. Available from: https://api.semanticscholar.org/CorpusID:73655224.
16. Higgins JPT, Altman DG, Gøtzsche PC, Jüni P, Moher D, Oxman AD, et al. The Cochrane collaboration's tool for assessing risk of bias in randomised trials. BMJ. 2011;343(7829):d5928.
17. Peterson J, Welch V, Losos M, Tugwell P. The Newcastle-Ottawa scale (NOS) for assessing the quality of nonrandomised studies in meta-analyses. Ottawa Ottawa Hosp Res Inst. 2011;2(1):1–12.
18. Guyatt GH, Oxman AD, Vist GE, Kunz R, Falck-Ytter Y, Alonso-Coello P, et al. GRADE: an emerging consensus on rating quality of evidence and strength of recommendations. BMJ [Internet]. 2008;336(7650):924–6. Available from: https://www.bmj.com/lookup/doi/10.1136/bmj.39489.470347.AD
19. How a Meta-Analysis Works. Introduction to Meta-Analysis [Internet]; 2009. p. 1–7. Available from: https://doi.org/10.1002/9780470743386.ch1.
20. Khoshdel A, Attia J, Carney SL. Basic concepts in meta-analysis: a primer for clinicians. Int J Clin Pract [Internet]. 2006;60(10):1287–94. Available from: https://onlinelibrary.wiley.com/doi/10.1111/j.1742-1241.2006.01078.x.

Meta-analysis Explanation and Guidance

Ahmed Saad Elsaeidy, Reem Sayad,
Rahma Sameh Shaheen, Ahmed M. Kedwany,
and Cyrus Yazdi

> *In God we trust. All others must bring data*
>
> —W. Edwards Deming

Abbreviations

ANCOVA	Analysis of covariance
CI	Confidence interval
GIVM	Generic inverse variance method
HR	Hazard ratio
IQR	Interquartile range
LVEF	Left ventricular ejection fraction
MD	Mean difference
OR	Odds ratio
PICO	P: Patient.; I: Intervention.; C: Comparison.; O Outcome.

A. S. Elsaeidy (✉) · R. S. Shaheen
Faculty of Medicine, Banha University, Banha, Egypt
e-mail: Ahmed170132@fmed.bu.edu.eg; rahma193226@fmed.bu.edu.eg

R. Sayad
Department of Histology, Faculty of Medicine, Assiut University, Assiut, Egypt
e-mail: reem.17289806@med.aun.edu.eg

A. M. Kedwany
Faculty of Medicine, Assiut University, Assiut, Egypt
e-mail: Ahmed.18313358@med.aun.edu.eg

C. Yazdi
Department of Anesthesia, Critical Care, and Pain Medicine, Beth Israel Deaconess Medical Center, Harvard Medical School, Boston, MA, USA
e-mail: cyazdi@bidmc.harvard.edu

© The Author(s), under exclusive license to Springer Nature Switzerland AG 2025
I. M. Shehata, O. Viswanath (eds.), *How to Successfully Publish a Manuscript*,
https://doi.org/10.1007/978-3-031-92538-2_15

PRISMA	Preferred Reporting Items for Systematic Reviews and Meta-Analysis
RCTs	Randomized clinical trials
RD	Risk difference
RevMan	Review Manager Software
RR	Relative risk, Risk ratio
SD	Standard deviation
SDs	Standard deviations
SE	Standard error
SEM	Standard error of the mean
SMD	Standardized mean difference

15.1 Introduction

– **Background and Definition**

The volume of information produced in medical research is becoming increasingly overwhelming, even for experienced researchers. With new studies being continuously published, clinicians find it nearly impossible to stay up-to-date, even within their own specialties. Moreover, individual studies frequently fail to provide definitive answers. [1]

To better understand this vast amount of information, review articles that combine the results of multiple studies are becoming increasingly common. When these reviews follow specific principles and the data are quantitatively analyzed, they are known as meta-analyses. A PubMed search for "Meta-Analysis" [Publication Type] yielded 209,922 results on Nov 8, 2024. Meta-analysis is an objective and quantitative method used to synthesize research findings. This study design integrates multiple research papers to evaluate the impacts of various interventions and strategies on different outcomes [2].

– **Importance**

Meta-analyses are among the most reliable forms of evidence in evidence-based medicine, as they offer less biased and more precise estimates on clinical issues [2]. By combining results from previous studies, meta-analyses enhance statistical strength and precision, addressing issues related to small sample sizes, rare outcomes, and weak statistical power [3]. Meta-analyses can identify sources of heterogeneity across studies, revealing subgroups (e.g., dependent on characteristics including sex, age, or duration of follow-up) where an intervention may be particularly beneficial or harmful. By examining these variations, meta-analyses can also highlight areas where further research is needed, helping to generate new hypotheses and guide the design of future studies. [4]

Meta-analyses also improve the precision of effect estimates in comparison to individual studies and help in the generalization of study outcomes. While the outcomes of a single study might be confined to its specific population, similar effects observed across various studies suggest broader applicability [5, 6].

- **PRISMA Guidelines**

The PRISMA (Preferred Reporting Items for Systematic Reviews and Meta-Analyses) statement mandates that authors thoroughly describe all data sources, including databases and their coverage dates. The authors must also specify the dates when they searched the databases and detail the search strategy they used to ensure the process can be replicated. [7, 8]

- **Aim of This Chapter**

In this chapter, we highlight several key studies as primary examples. Readers will gain the most benefit by reviewing these studies in conjunction with this chapter. An overview of these studies is presented in Table 15.1.

This chapter aims to provide a thorough guide to conducting meta-analyses, covering fundamental concepts, data management, handling missing data, and various statistical methods for different outcomes. It also addresses result presentation, heterogeneity, and advanced topics like sensitivity analyses and meta-regression to equip researchers with essential tools and knowledge.

Table 15.1 Table of example studies used in this chapter

Study ID	Title	Number of included studies	Sample size	Statistical analysis methods
Elsaeidy et al. (2024)	Efficacy and safety of extracorporeal membrane oxygenation for cardiogenic shock complicating myocardial infarction: a systematic review and meta-analysis	4 RCTs	611	Statistical analysis was conducted using RevMan V5.4 with a fixed-effect model. Risk ratios and 95% confidence intervals were calculated, with $P < 0.05$ considered significant. Heterogeneity was assessed using chi-square ($\alpha < 0.1$) and I^2 tests, following Cochrane guidelines. Trial sequential analysis was performed to ensure reliability
Shaheen et al. 2024	Efficacy and safety of 12-h versus 24-h magnesium sulfate in the management of patients with pre-eclampsia and eclampsia: a systematic review and meta-analysis	13 RCTs	2813	Statistical analysis was conducted using RevMan V5.4 with a fixed-effect model. Risk differences and 95% confidence intervals were calculated, with $P < 0.05$ considered significant. Heterogeneity was assessed using chi-square ($\alpha < 0.1$) and I^2 tests, following Cochrane guidelines

(continued)

Table 15.1 (continued)

Study ID	Title	Number of included studies	Sample size	Statistical analysis methods
Elsaeidy et al. (2024)	Efficacy and safety of ketamine-dexmedetomidine versus ketamine–propofol combination for periprocedural sedation: a systematic review and meta-analysis	22 RCTs	1429	Continuous and dichotomous data were pooled as mean difference (MD) and odds ratio (OR) with 95% confidence intervals using the inverse-variance method and a random-effects model. Heterogeneity was assessed with chi-square ($\alpha < 0.1$) and I^2 tests, interpreted following Cochrane guidelines
Elsaeidy et al. (2024)	The efficacy and safety of levosimendan in patients with advanced heart failure: an updated meta-analysis of randomized controlled trials	15 RCTs	1181	Meta-analysis was conducted using RevMan with a random-effects model (DerSimonian and Laird method). Risk ratios (RR) and mean differences (MD) with 95% CIs were calculated. Heterogeneity was assessed with chi-square and I^2 tests. Leave-one-out sensitivity analysis and trial sequential analysis were performed to ensure reliability

15.2 Preparation for Meta-analysis

A systematic review is a prerequisite for conducting a meta-analysis, which makes robust systematic review methods essential (**see the Systematic Review chapter**). Conducting a systematic review includes many key steps, outlined as follows: formulating a detailed research question, detailing a search strategy, establishing detailed inclusion and exclusion criteria, utilizing multiple academic databases (such as PubMed, Medline, Embase, Scopus, and Cochrane Library); then, the studies are independently screened, followed by data extraction and quality assessment of the included studies. Misconducting any of these steps can compromise the meta-analysis quality [9].

15.2.1 Data Management and Analysis Planning

Study data should be extracted onto pre-prepared extraction sheets, with careful consideration given to the data needed before starting the review to avoid missing any crucial information. Ideally, data extraction should be performed with a double-check approach to identify transcription errors and minimize subjectivity, especially when data formats differ from those required on the extraction sheets [10]. It is

15 Meta-analysis Explanation and Guidance

recommended to establish analytic methods prior to data extraction and analysis. However, this decision is influenced by the reported effect measures and the study designs of included studies. When reviewers seek to determine the mean difference (MD) or standardized mean difference (SMD), they must gather study-level means, standard deviations (SDs), and sample sizes for both the intervention and control groups [11].

Note that
- In comparative studies, when outcomes are measured using the same units, MD between experimental and control groups can be aggregated to give an overall estimate. If different measures are used across studies, the SMD is calculated by dividing each MD by their respective SDs, allowing for comparison across studies. [12]
- Inconsistencies in effect size units, such as blood pressure reported in different units (e.g., mmHg, kPa), need to be standardized for comparison across studies. This necessitates extracting the necessary experimental parameters from the included studies. Alternatively, using the SMD can facilitate comparisons across studies despite these unit differences. [11]

15.2.1.1 Effect Size Measure Choice and Conversion

Measures of effect size are statistical constructs used to assess the results between two intervention groups. For example, **the risk ratio (RR)** measures the probability of an event happening in the treatment group relative to its probability in the control group. [13]

Choosing effect size measures can vary depending on the subfield of investigation [14]. The absolute effect size, calculated as an absolute difference from the baseline, is straightforward and contextually informative but requires a standard scale or conversion metrics [15]. When establishing a standard scale is difficult, scale-free measures such as standardized or relative metrics are employed [16].

Once the primary effect size measure is defined, it may be essential to convert differently reported study findings to the chosen primary measure. Conversion formulas provided by Borenstein et al. (2009) or Lipsey and Wilson (2001) can be used to harmonize different effect size measures. Additionally, online effect size calculators are available to assist with these conversions, as the tool discussed in the next paragraph [17, 18].

15.2.1.2 New Tool for Statistical Data Conversion

What Are the Challenges?
As previously mentioned, meta-analysis synthesizes findings from various studies, providing strong insights into intervention effects and supporting evidence-based medicine. Nevertheless, this can be hindered by issues including incoherent reporting of data, complicated calculations, and limited time. [19] Authors face the overwhelming task of standardizing different statistical measures, which can be both labor-intensive and prone to error without suitable tools. [20]

How to Overcome Them?
To handle these challenges, the Meta-Analysis Accelerator was created [21]. This tool provides 21 distinct statistical conversions, such as transforming median and interquartile range (IQR) into mean and SD, converting the standard error of the mean (SEM) into SD, also translating confidence intervals (CI) into SD. Launched in 2023 as Meta Converter and later rebranded as Meta-Analysis Accelerator, it has promptly garnered global interest, attracting over 12,200 visits from countries like Egypt, France, Indonesia, and the USA between March and May 2024. This tool can be accessed through the link https://meta-converter.com/.

What Are the Advantages of This Tool?
Designed with a user-friendly interface, the tool allows for simple navigation and efficient, accurate conversions. By significantly improving the precision of meta-analyses, the Meta-Analysis Accelerator empowers researchers to concentrate on data interpretation rather than manual computations, thereby enhancing the quality and accessibility of meta-analyses. [22]

15.2.1.3 Weighting Schemes
A key characteristic of meta-analysis is the capacity to assign greater weight to larger, more thoroughly reported studies, enhancing the quality and reliability of the results. This is typically done by the generic inverse-variance method (GIVM) that combines total variance and sample size, giving more significance to studies having larger sample sizes and fewer experimental errors. However, this method is valid only if the sampling error is random, the reported effects exhibit consistent variance, and the sample size corresponds to the number of independent observations. If these conditions are not met, weighting based on sample size can serve as an alternative method. [11]

15.2.2 Missing Data

15.2.2.1 Types of Missing Data
Missing data can arise from various sources in systematic reviews or meta-analyses. These include entire studies, specific outcomes within studies, summary data for outcomes, and individual participant data.

Sources of Missing Data
- **Entire studies** may be absent from a review due to nonpublication, obscure publication venues, low citation rates, or improper indexing in databases [23]. Review authors must remain vigilant about the risk of overlooking pertinent studies, particularly due to publication bias, which often results in the nonpublication of studies with unfavorable outcomes.
- Certain studies might not provide data on key **outcomes**, such as patient satisfaction or long-term complications. It can be challenging to establish if these outcomes were not measured or simply not reported, which may be influenced by selective outcome reporting bias.

- Essential **summary data** for outcomes, such as SDs for continuous outcomes, sample sizes, event counts, SE, follow-up durations, and details for time-to-event outcomes, may be missing. [23, 24] This is particularly problematic for change-from-baseline outcomes. Inappropriate analyses of certain study designs, like cluster-randomized and crossover trials, can also result in missing summary data. At times, approximations can be made, such as ascribing correlation coefficients or SDs. [25]
- In many studies, some **participants** may be missing from the reported results. This can also impact subgroup analyses and meta-regressions (see Sects. 15.5.3.2 and 15.5.3.3.), which require detailed study-level characteristics to distinguish between studies.

By understanding and addressing these possible sources of missing data, authors can enhance the robustness and accuracy of their meta-analyses. It is crucial to consider the reasons behind missing data. Statisticians categorize missing data into two main types:

- *Missing at random*

Data are deemed "missing at random" if the absence of data is not relevant to the real values of the missing data. For example, if some participants forget to return their quality-of-life questionnaires, this omission is likely unrelated to their actual quality of life. Analyses based on "missing at random" data are generally unbiased but may involve a reduced sample size. [26]

- *Not missing at random*

Data are deemed "not missing at random" when the absence of data is relevant to the actual missing values. As an example, in a study on chronic pain, patients experiencing severe pain might be less likely to complete follow-up surveys, which results in missing outcome data. Data like this cannot be ignored because meta-analyses depending solely on the existing data are likely to be problematic [23, 26].

15.2.2.2 General Principles for Dealing with Missing Data (1–4 Approaches)

Numerous studies have explored various techniques for handling missing data. In this section, we outline four fundamental principles from the Cochrane Statistical Methods Group:

- **Analyzing available data**: This approach involves using only the data that is available, effectively ignoring any missing data.
- **Substituting missing data** with replacement values and using these substitutes as though they were real observations. Common techniques include carrying forward the last observation, assuming all missing data indicate poor outcomes, imputing the mean, and using predicted values derived from regression analysis. This approach is practical and widely used, but it cannot consider the uncertainty linked to the imputed values, often resulting in an overly narrow CI.

- **Imputing missing data with uncertainty adjustment**: This involves imputing missing data while considering the uncertainty of these imputations. Techniques include multiple imputation and simple imputation methods with adjustments to the standard error (SE).
- **Using statistical models**, incorporating assumptions on how the missing data relate to the observed data.

The third and fourth approaches, which involve more sophisticated statistical techniques, typically require the expertise of a knowledgeable statistician. [27, 28]

15.2.2.3 Dealing with Missing Outcome Data from Individual Participants

Researchers can perform sensitivity analyses to evaluate the possible effects of missing outcome data. This process involves making assumptions about how the missing data relate to their true values. Many approaches can be employed for this purpose [29].

- Higgins et al. (2008) propose a strategy for managing missing dichotomous outcomes by adjusting assumptions about the event risk among missing participants relative to those with observed data while also considering the uncertainty these assumptions introduce. [30]
- Ebrahim et al. (2013) developed a comprehensive method to evaluate the effect of missing continuous outcome data on the risk of bias in trials using the same measurement instrument. Their approach begins with a complete case analysis and is followed by sensitivity analyses. Their study provides detailed guidance on how to perform these analyses. The extent to which results vary with these sensitivity analyses indicates the level of bias risk resulting from missing data. [31]
- Ebrahim et al. (2014) extended their methodology to meta-analyses which incorporate trials that use various instruments to measure the same outcome. [32]

Avoiding double-counting of events is essential, as trial reports often do not clarify if the reported event numbers include the entire randomized sample or only those participants who completed the study [33]. While "worst case" and "best case" analyses are traditionally used to define the extreme boundaries of possible outcomes, they may not be informative for the most likely scenarios [30].

15.3 Data Analysis and Presentation of Results

Meta-analysis can incorporate data on the quality and reliability of the original studies, which is a significant characteristic. This is accomplished by giving larger, more comprehensively documented studies more weight. The total estimate and the measure of this estimate's variability are the most important outcomes of the meta-analysis. [11]

The following are the different data types of meta-analysis:

15.3.1 Generic Inverse-Variance Method (GIVM)

An extremely popular and basic variation of the meta-analysis process is known as the GIVM. This method applies to various situations encountered by authors. [23]

The GIVM requires an estimate of the treatment's relative effect and SE for each study in the analysis.

- The estimation of the relative treatment effect of each included study has a certain weight equal to the variance of the effect estimate (i.e., 1/SE [2]).

Thus, larger trials, characterized by low SE, are afforded greater significance than smaller studies, which exhibit higher SE. This selection of weights reduces the imprecision (uncertainty) of the aggregated effect estimate [23, 34].

This method should be employed only when the entry of data in the conventional formats of dichotomous, continuous, or individual patient data is unfeasible, to guarantee that the reader can view the data categorized by the treatment group whenever possible. Moreover, GIVM enables the examination of well-analyzed nonrandomized trials, crossover trials, and cluster-randomized trials in addition to outcome data that are rates, time-to-event, or ordinal [23].

- **N.B.:** The GIVM should not serve as a substitutional method for the other methods already available in RevMan 4.1 but rather in circumstances when those methods are not applicable.

There are two scenarios in which authors may need to employ the GIVM:

1. **When the result is dichotomous:** GIVM can assume only one of two possibilities: the published study presents only the odds ratio (OR) or RR together with its SE. In the old version of RevMan (4.1), such outcomes were excluded from quantitative analysis and could solely be articulated in the review's text.
   - **Example:** There is an outcome reported by five studies, and two of them reported only the RR, OR, and SE without reporting the number of participants that had this outcome in the intervention and control groups. These two studies in the old version of RevMan 4.1 will not be included in the analysis, but now we can incorporate them by using GIVM [35].
2. **When the outcome is continuous**: GIVM should be used when the published paper solely presents the difference between the means of the two groups and the SE of this difference [35].

The following are the steps needed to add the data into RevMan in the form of GIVM:

1. Add New Comparison (Fig. 15.1).
2. In the New Comparison Wizard, enter the name of your comparison (Fig. 15.2).
3. Add a new outcome (Fig. 15.3).

4. In the New Outcome Wizard window, choose "Generic Inverse Variance" (Fig. 15.4). Then, press next.
5. In the name of the outcome, enter the name of your outcome (Fig. 15.5), and label each arm of the comparison.
6. The next step is choosing the analysis methods according to the characteristics of the data (Fig. 15.6).

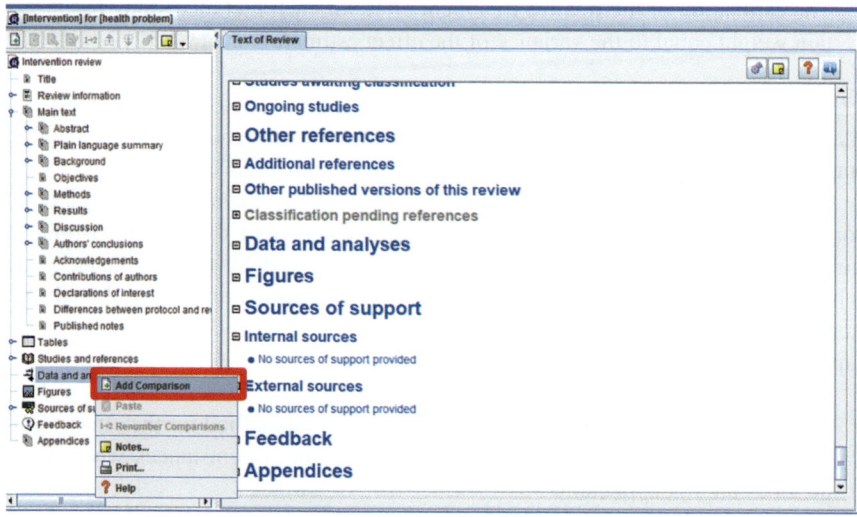

Fig. 15.1 Add new comparison in RevMan

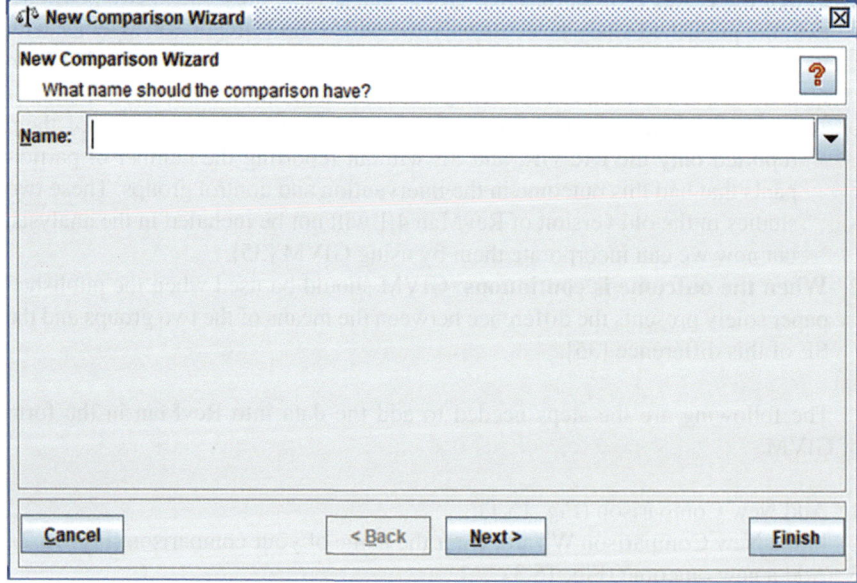

Fig. 15.2 Defining comparison labels in RevMan

15 Meta-analysis Explanation and Guidance

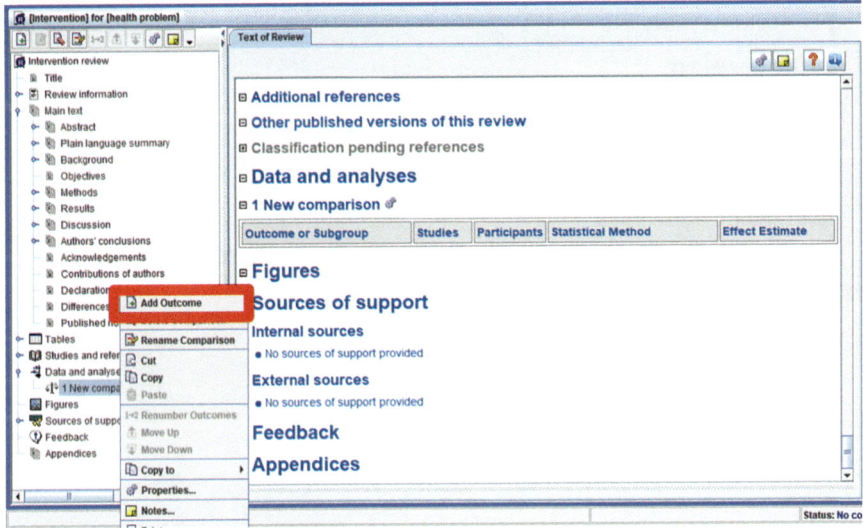

Fig. 15.3 Add new outcome in RevMan

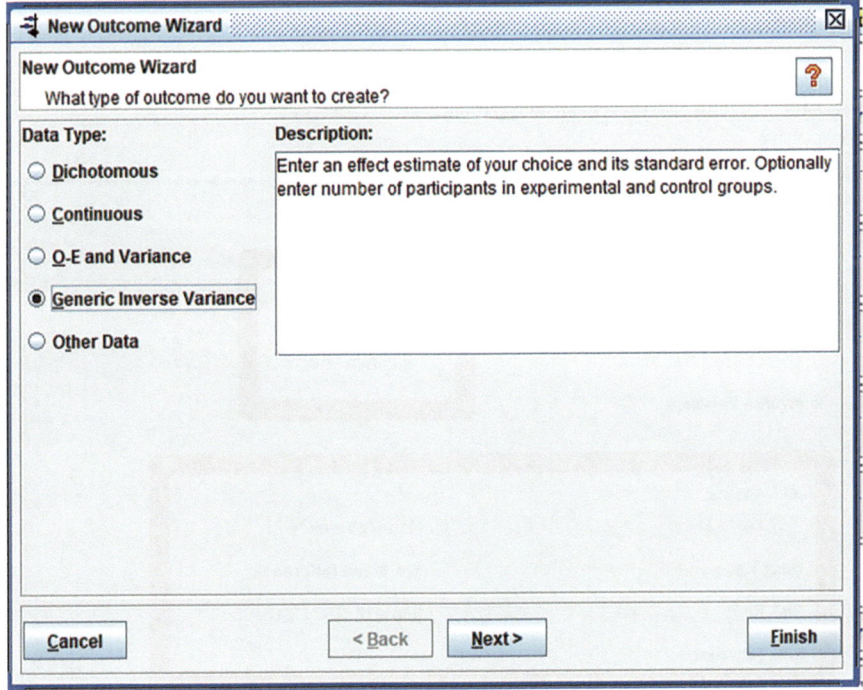

Fig. 15.4 Choosing the "generic inverse variance" method in RevMan

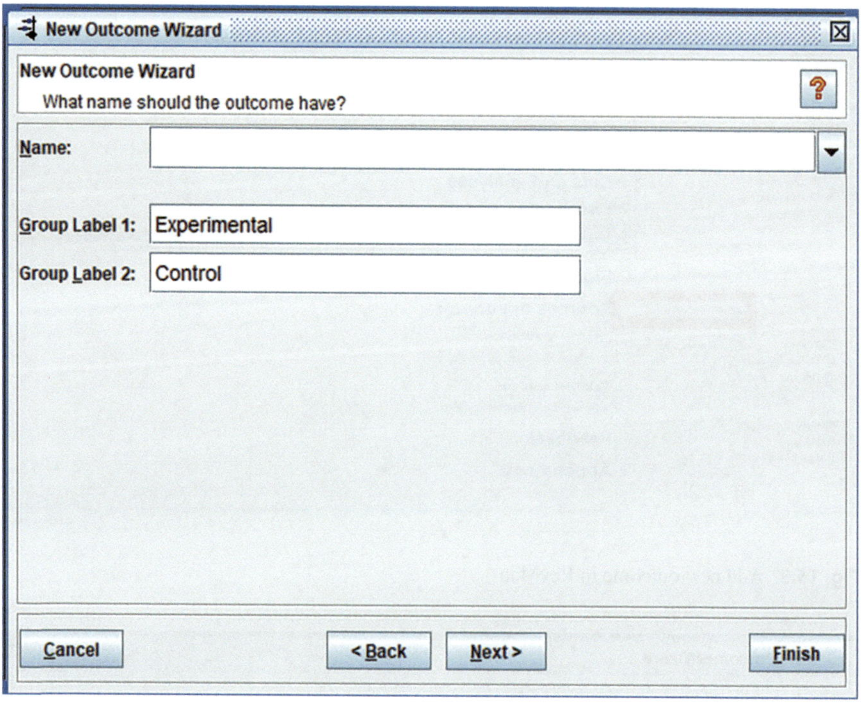

Fig. 15.5 Defining outcome labels in RevMan

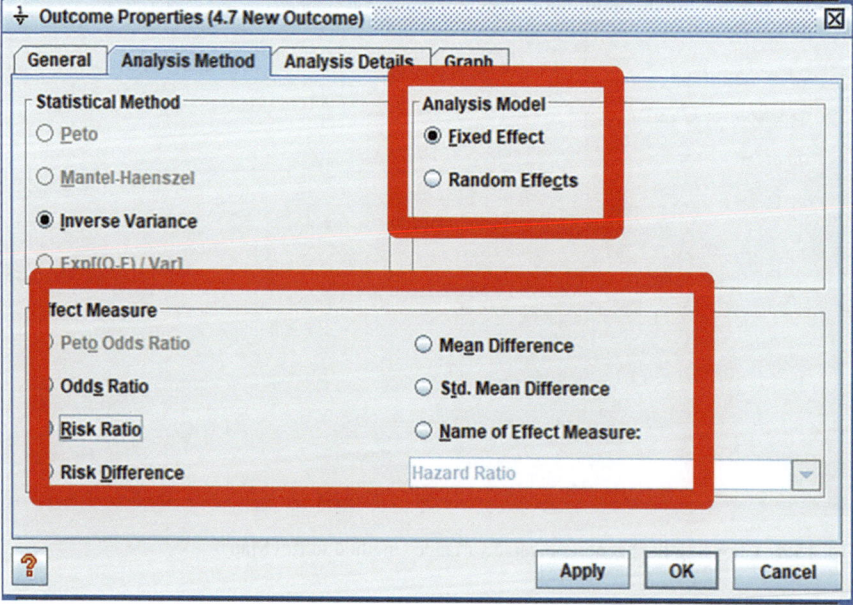

Fig. 15.6 Choosing the analysis methods in RevMan

15 Meta-analysis Explanation and Guidance

Table 15.2 Comparison between fixed- and random-effect models

Fixed-effect model [36–38].	Random-effect model [11, 39–41]
One true effect	True effect varies
Differences between studies due to sampling error only	Differences between studies more than sampling error
Accounts for within-study variations	Accounts for within- and between-study variations
The null hypothesis for heterogeneity is that the common effect size is (for example) zero	The null hypothesis for heterogeneity is that every study in a meta-analysis has the same underlying effect size
Summary effect = common effect	Summary effect = mean effect
The summary effect is limited to the identified group	The summary effect is more generalizable

Table 15.3 Outcomes of a two-group randomized experiment with a binary result can be represented in a 2×2 table

	Event ('Success')	No event ('Fail')	Total
Experimental intervention	S_E	F_E	N_E
Comparator intervention	S_C	F_C	N_C

Abbreviations: S_E Number of participants that have events in the experimental group, F_E Number of participants that didn't have events in the experimental group, N_E Number of participants in the experimental group, S_C Number of participants that have events in the control group, F_C Number of participants that did not have events in the control group, N_C Number of participants in the control group

Each type of these analysis methods will be discussed in this chapter (Sects. 15.3.1.1 and 15.3.1.2.).

15.3.1.1 Fixed and Random-Effect Methods for Meta-analysis

The random- and fixed-effect models are prominent statistical models of meta-analysis. The choice of the model is of crucial significance. The model not only influences the computations but also delineates the study's objectives and the interpretation of the results.

The following table has the most important differences between fixed- and random-effect models (Table 15.2).

15.3.1.2 Conducting Meta-analyses Using Inverse Variance

Inverse-variance meta-analyses are carried out by the majority of meta-analysis systems. When the result is continuous, it is typical for users to provide sample sizes, means, and SDs for each group. Users frequently provide summary data from each intervention arm of each trial, such as a 2 x 2 table, where the outcome is binary (Table 15.3) [13].

When each intervention group's summary statistics are available, you can immediately apply the inverse-variance method. For instance, you can immediately enter estimates and associated SE into RevMan under the GIVM result type. The log OR and the SE of the log OR are examples of natural logarithms that must be entered into RevMan in order to calculate the ratio measurements of the intervention effect

[23]. This option greatly enhances the meta-analysis's flexibility by allowing the entry of estimates and SE. It makes the condition easier to assess properly analyzed nonrandomized trials, crossover trials, and cluster-randomized trials [42].

15.3.2 Meta-analysis of Dichotomous Outcomes

The risk difference (RD), risk ratio (relative risk-RR), and OR are used to measure the effect between the dichotomous data. Each of these is a valid metric. While RR is simpler to interpret, OR has a purer mathematical foundation. CI (often 95%, but adjustable) is provided for each metric [23, 35].

There are four common approaches to meta-analysis for dichotomous outcomes:

- Fixed-effect models
  - Inverse-variance methods
  - Mantel-Haenszel methods
  - Peto methods
- Random-effect model
  - DerSimonian and Laird inverse variance

RevMan offers all of these methods as analytical options. Only OR can be combined using the Peto method, while the other three methods can combine OR, RR, or RD [23].

- **Be aware** that the absence of events in a group, often referred to as "zero cells," presents challenges in the calculation of estimates. A specific section will discuss SE using certain methods [23]. Introduce the following methods.

15.3.2.1 Mantel-Haenszel Methods
The estimations of SE for effect estimates used in the inverse-variance method may not be sufficient when data are limited, either because of small sample sizes or low event risks. A unique weighting strategy based on the particular effect measure (e.g., RR, OR, and RD) used the Mantel-Haenszel methods, which are fixed-effect meta-analysis methods [43, 44]. They exhibit superior statistical features in scenarios with limited events. Fixed-effect meta-analyses usually employ the Mantel-Haenszel method instead of the inverse-variance method. In other situations, the estimations of both methods are comparable [23, 43].

15.3.2.2 Peto Odds Ratio Method
Peto's method is exclusively applicable to the aggregation of OR [45]. It employs an inverse-variance methodology, utilizing an approximate technique for computing the log OR and applying distinct weights. A different way to look at the Peto method is as a summation of "O–E" statistics. Here, O stands for the observed number of events and E for the expected number of events in each study's experimental intervention group, assuming that there is no intervention effect [23, 46].

The method used in calculating the log OR is effective when intervention effects are minimal (OR are around 1), incidents are relatively low, and the studies have comparable sample sizes in both experimental and comparator groups. In other instances, it has demonstrated a propensity to provide biased responses. It is not advised to use Peto's method as a standard methodology for meta-analysis due to the inconsistent fulfilment of these criteria.

The Peto method does not require adjusting zero cell counts. This strategy is effective for infrequent events [46]. Peto's technique can also be employed to integrate research with dichotomous outcome data alongside studies utilizing time-to-event analyses that have applied log-rank tests, as will be illustrated in a specific section.

We enter a dichotomous outcome by the following steps:

- All steps are the same as the previous steps except step 4 and step 6 (Table 15.4).
- Step 4: In the New Outcome Wizard window, choose "Dichotomous" (Fig. 15.7).
- Step 6: The next step is choosing the analysis methods according to the characteristics of the data.
    a: If the statistical method is "Peto," the analysis model will be "fixed-effect," and the effect measure will be "Peto odds ratio" (Fig. 15.8).

Table 15.4 Data analysis by RevMan

Step 1	Add New Comparison.	Fig. 15.1
Step 2	In the new comparison wizard, enter the name of your comparison.	Fig. 15.2
Step 3	Add a new outcome.	Fig. 15.3
Step 4*	In the new outcome wizard window, choose the "data type." Then, press next.	Fig. 15.4
Step 5	In the name of the outcome, enter the name of your outcome and label each arm of the comparison.	Fig. 15.5
Step 6#	The next step is choosing the analysis methods according to the characteristics of the data.	Fig. 15.6
Step 7	Through "outcome properties," select "graph" and then in the "Sort by" section.	Fig. 15.16
Optional steps		
Step 8	In the new outcome wizard window, select "add a subgroup analysis for the new outcome.	Fig. 15.23
Step 9	In the new subgroup wizard window, write the name of the subgroup.	Fig. 15.24
Step 10	In the new subgroup wizard window, choose "add study data for the new subgroup.	Fig. 15.25
Step 11	In the new study data wizard window, select the included studies in the outcome of interest. After choosing them, select finish.	Fig. 15.26

(*) Step 4 is varied according to the type of data, e.g., Figs. 15.4, 15.7, 15.11, or 15.13
(#) Step 6 is varied according to analysis methods, e.g., Figs. 15.6, 15.8, 15.9, 15.10, 15.12, or 15.14

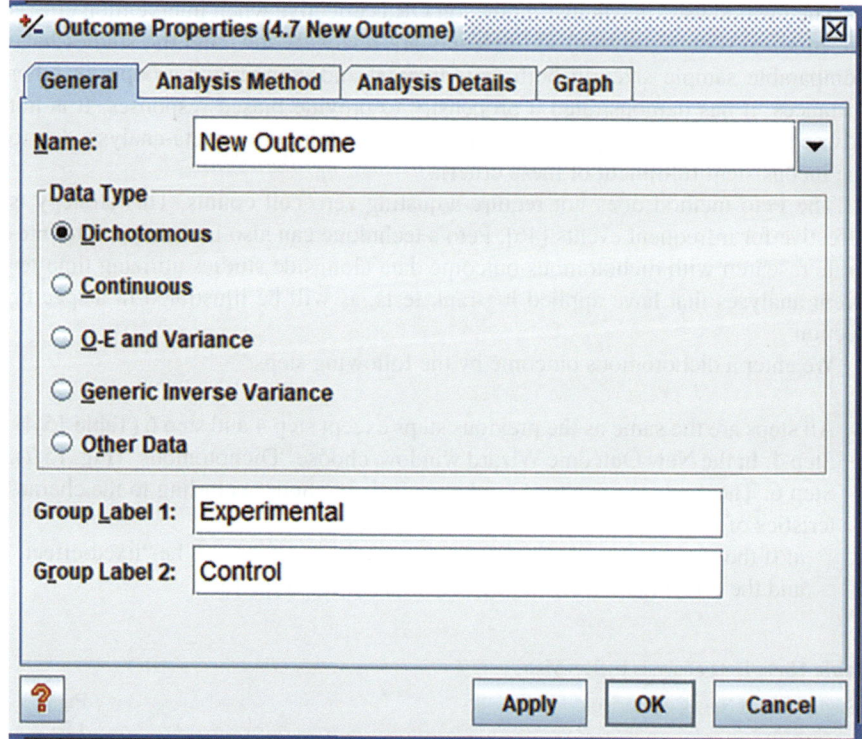

Fig. 15.7 Choosing the "Dichotomous" method in RevMan

6. b: If the statistical method is "Mantel-Haenszel," the analysis model will be a "fixed-effect" model, and the effect measure will be "odds ratio, risk ratio, or risk difference" (Fig. 15.9).

6. c: If we choose "inverse variance" for the statistical methods, then we will choose the "fixed-effect" model. According to the effect measure, we can choose "odds ratio, risk ratio, or risk difference" (Fig. 15.10).

Choosing which effect measure for the dichotomous outcomes will be discussed in the next section.

15.3.2.3 Which Effect Measure for Dichotomous Outcomes?

The impact of an intervention can be presented as either a relative effect or an absolute effect. RR and OR represent the relative metrics, but RD represents the absolute metrics. Additionally, there exist two distinct risk ratios. We can calculate event occurrence and non-occurrence RRs. These yield varying summary outcomes in a meta-analysis, occasionally in a dramatic manner [13, 23].

The choice of a summary statistic for meta-analysis hinges on the equilibrium of three criteria [47]:

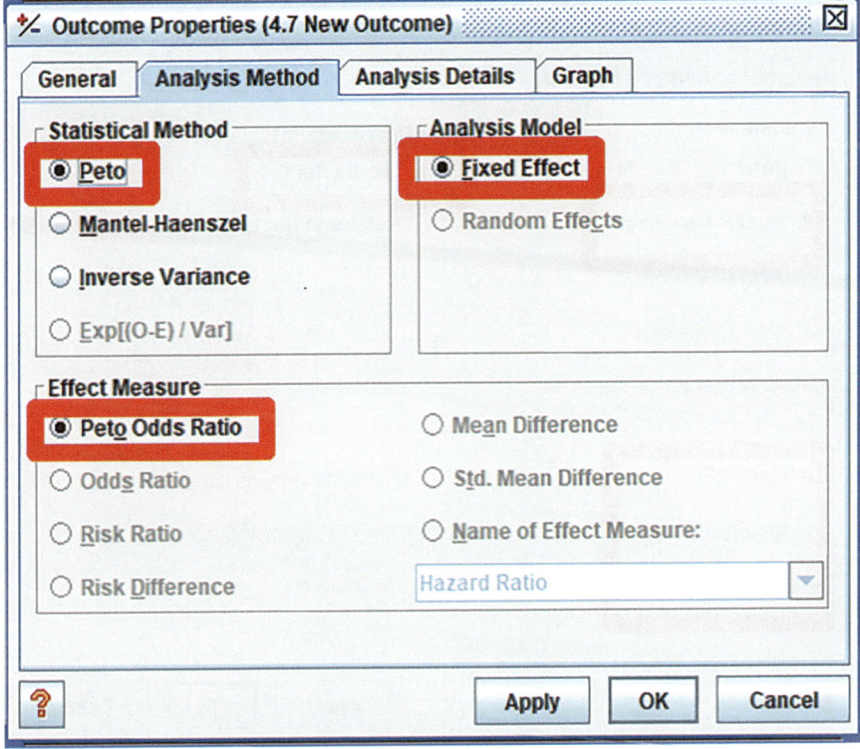

Fig. 15.8 Choosing the "Peto" statistical method in RevMan

1. We aim for a summary statistic that produces comparable values of all meta-analyses' included studies and the demographic subgroups that will receive the therapies. It is more realistic to present the effect of the intervention as a single summary value when the summary statistic is more consistent.
2. They need to have specific characteristics of mathematics to carry out a noteworthy meta-analysis.
3. They should be readily comprehensible and applicable to the review's readers. To effectively facilitate readers' interpretation and use of the results, present the summary intervention effect. There is no specific effect measure for dichotomous data that is universally superior; hence, the selection necessitates a compromise.

Relative effect estimates are typically more consistent than absolute measures, according to uniformity empirical data [47–49]. Therefore, it is prudent to refrain from doing meta-analyses of RD unless there is a compelling rationale to anticipate consistency in RD within a certain clinical context. Generally, the OR and RR exhibit minimal disparity regarding consistency [47]. When the study seeks to diminish the occurrence of an unfavorable event, empirical evidence indicates that

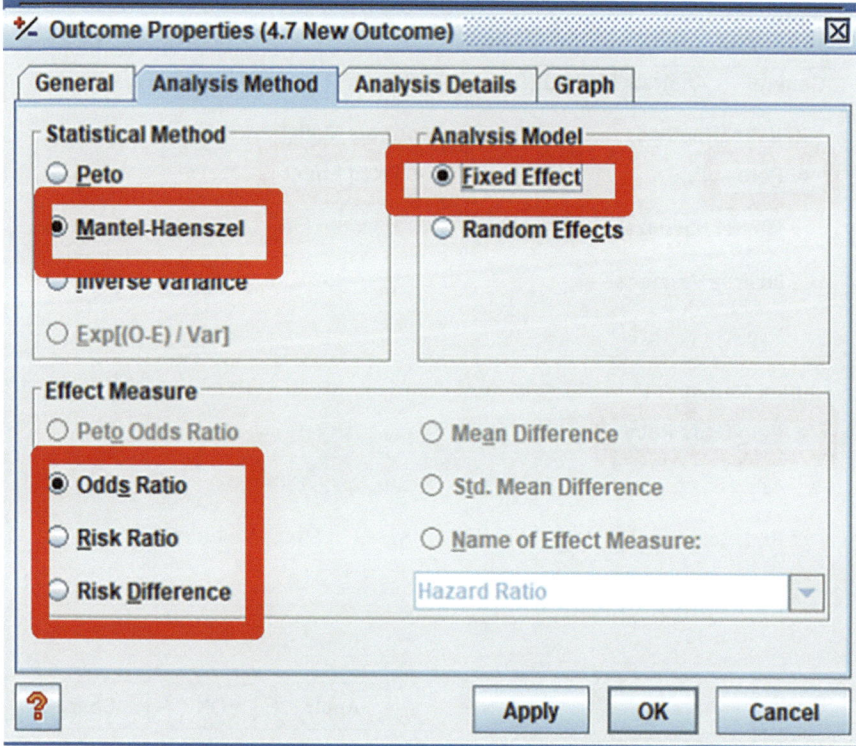

Fig. 15.9 Choosing the "Mantel-Haenszel" statistical method in RevMan

RR for the adverse event exhibits greater consistency than those for the nonevent [47]. It is typically not advisable to choose an effect measure based on its consistency in a specific context, as this could lead to a selection that falsely enhances the precision of a meta-analysis estimate.

The OR is the most challenging summary statistic to comprehend and implement in practice, with numerous doctors indicating difficulty in its application. Numerous reported instances exist in which authors have improperly interpreted OR from meta-analyses as RR. Although it is possible to reframe OR for interpretation, there is a legitimate worry that consistently presenting systematic review outcomes as OR could lead to an overestimation of the benefits and harms of interventions in clinical practice [50]. On the other hand, absolute measurements of effect are considered more interpretable by clinicians than relative effects [50], facilitating the assessment of trade-offs between potential benefits and hazards of therapies. Nevertheless, their generalizability is diminished [23].

Meta-analyses are typically advised to utilize RR (ensuring a judicious selection of the outcome category designated as the event) or OR. This is due to the necessity of avoiding summary statistics that empirical evidence suggests are unlikely to produce reliable intervention effect estimates (the RD), and it is infeasible to utilize

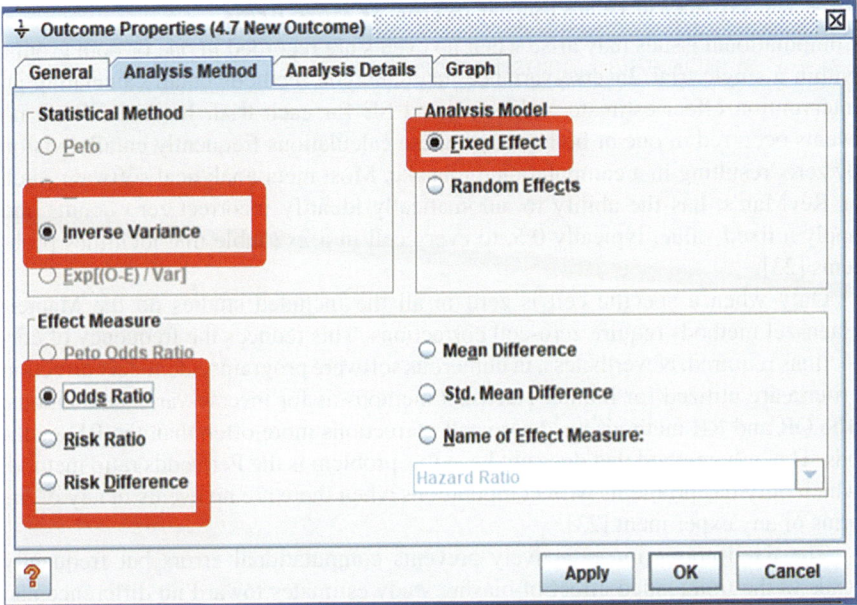

Fig. 15.10 Choosing the analysis method according to the statistical method of dichotomous outcome in RevMan

statistics for which meta-analysis cannot be conducted. It might be prudent to do a sensitivity analysis to determine if the selection of summary statistic and event category significantly impacts the outcomes of the meta-analysis [13, 23].

15.3.3 Meta-analysis of Rare Events

There is no definitive risk threshold that categorizes incidents as "rare." Risks of **1 in 1000** are undoubtedly considered rare events, and numerous individuals would categorize risks of **1 in 100** similarly. Nevertheless, the concerns addressed in this section may also influence the efficacy of methods when risks reach **1 in 10**. On the other hand, we can consider the outcome to be rare when a significant percentage of the studies in the meta-analysis report no occurrences in one or more study groups [23].

When it comes to rare outcomes, meta-analysis might be the only trustworthy method to gather data regarding the effectiveness of healthcare interventions. Single studies sometimes lack sufficient power to identify differences in rare outcomes; however, a meta-analysis encompassing multiple studies may possess the power to assess the influence of interventions on the occurrence of the uncommon event. Nevertheless, several meta-analytic techniques rely on large sample approximations and are inappropriate for instances of infrequent events. So, authors must be careful when choosing a meta-analysis approach [51].

15.3.3.1 Studies with No Events in One or More Arms

Computational issues may arise when no events are recorded in one or both groups within a single trial. Inverse-variance meta-analytic methods entail calculating an intervention effect estimate and associated SE for each trial. In cases where no events occurred in one or both groups, these calculations frequently entail division by zero, resulting in a computational mistake. Most meta-analytical software, such as RevMan's, has the ability to automatically identify incorrect zero counts and apply a fixed value, typically 0.5, to every cell in a 2x2 table that identifies problems [23].

Only when a specific cell is zero in all the included studies do the Mantel-Haenszel methods require zero-cell corrections. This reduces the frequency of corrections required. Nevertheless, in numerous software programs, identical correction criteria are utilized for Mantel-Haenszel methods as for inverse-variance methods. The OR and RR methods need zero cell corrections more often than the RD methods. The only method that does not have this problem is the Peto odds ratio method, which only has problems with computations when there are no events in any of the arms of any experiment [23].

The fixed correction effectively prevents computational errors but frequently leads to the unintended effect of biasing study estimates toward no difference and inflating their variances, thus improperly reducing their impact on the meta-analysis. Disparities in the sizes of study arms, more common in nonrandomized studies than in randomized trials, will result in a directional bias about the treatment effect. Sweeting et al. (2004) looked into a number of nonfixed zero-cell corrections. One correction was proportional to the inverse of the contrasting study arm size. They discovered that this approach proved more effective than fixed 0.5 corrections in cases where the arm sizes were not equal to [52].

15.3.3.2 Studies with No Events in Both Arms

We eliminate studies that report no incidents in both groups in the usual practice of meta-analysis of OR and RR. This is because such studies fail to provide both the direction and degree of the relative treatment impact. It is clear that events do not happen very often in either the comparator or the experimental intervention, but it is not clear which group may have a higher risk or whether the risks are similar or different in size (when risks are small, they can mean either substantial or negligible ratios). Although one might be inclined to deduce that the risk is minimized in the group with the larger sample size (due to a lower upper limit of the CI), this assumption is unfounded, as the sample size allocation was established by the study investigators and does not reflect the event's incidence [23].

RD methods possess an advantage over OR methods, as the RD can be defined as zero when no events transpire in either group. Therefore, the estimating process incorporates such a study. Bradburn et al. (2007) used simulations to show that all RD methods give too broad of a CI when events are rare, which means they do not have enough statistical power. This makes them unsuitable for meta-analysis of rare outcomes [53].

This is especially important when looking at outcomes related to treatment safety, since finding major adverse events correctly is a crucial part of making new medicines.

Outcomes for which no events transpire in either arm are likely omitted from reports of several randomized trials, hence excluding them from meta-analysis. It remains ambiguous when analyzing published results whether the omission of a specific adverse event indicates its absence or merely because it was not designated as a monitored endpoint. Reporting outcomes without events alters the results of RD meta-analyses. However, methods based on OR and RR naturally leave out this data, no matter how published they are [23].

15.3.3.3 Validity of Meta-analysis Methods for Rare Events

Simulation trials have shown that several meta-analytical methods can produce erroneous conclusions for rare events, a phenomenon anticipated due to their reliance on asymptotic statistical theory. Their performance has been evaluated as substandard due to biased results, overly broad confidence intervals, or inadequate statistical power to detect significant differences.

It looks like the right choices depend on the risk of the control group, how big of an effect the treatment is expected to have, and how many people are in the experimental group and the control group in each trial [23]. Estimates of risks and odds are nearly equal when events are rare, and we can read the outcomes of both as ratios of probability.

Bradburn et al. (2007) discovered that numerous frequently employed meta-analytical techniques exhibited bias in instances of rare events [53]. Inverse variance, DerSimonian and Laird odds ratio, and RD approaches, as well as the Mantel-Haenszel odds ratio method with a 0.5 zero-cell correction, showed the strongest bias. When the risk of events was low, as we previously stated, RD meta-analytical approaches had reduced statistical power and an excessively large CI [23].

The Peto one-step odds ratio method has been identified as the least biased and most potent method at event rates below 1%, yielding optimal CI coverage, contingent upon the absence of significant imbalance between treatment and comparator group sizes within studies and the treatment effects not being excessively large. Sweeting et al. (2004) consistently observed this finding across three distinct meta-analytical methods (Mantel-Haenszel, Peto, and inverse variance) [23, 52].

15.3.4 Meta-analysis of Continuous Outcomes

The MD or SMD are the presentation formats for continuous data. Each metric provides the CI, which is typically 95% but is adjustable. Most of the time, when you do a meta-analysis of continuous data, you assume that the results are spread out normally within each intervention group in each trial. This assumption may not consistently hold; however, it is inconsequential in extensive investigations. It is prudent to contemplate the potential for biased data [23, 35].

A continuous outcome is entered into the RevMan application by the following steps:

- **All steps a**re the same as the previous steps, except steps 4 and 6 (Table 15.4).
- Step 4: In the New Outcome Wizard window, choose "Continuous" (Fig. 15.11).
- Step 6: If the statistical method is "inverse variance," the analysis method will be a "fixed-effect" model, and the effect measure will be "mean difference or standardized mean difference" (Fig. 15.12).

Choosing which effect measure for the continuous outcomes will be discussed in the next section.

15.3.4.1 Which Effect Measures for Continuous Outcomes?

The two summary statistics frequently employed in meta-analysis of continuous data are the MD and SMD. Whether studies provide outcomes on a uniform scale (allowing for the use of MD) or on different scales (often necessitating the SMD) largely influence the choice of summary statistics for continuous data [54].

The distinct functions of SDs of outcomes in MD and SMD methodologies must be comprehended. In the MD method, the SDs are utilized alongside the sample sizes to calculate the weight assigned to each study. Studies with smaller SDs are

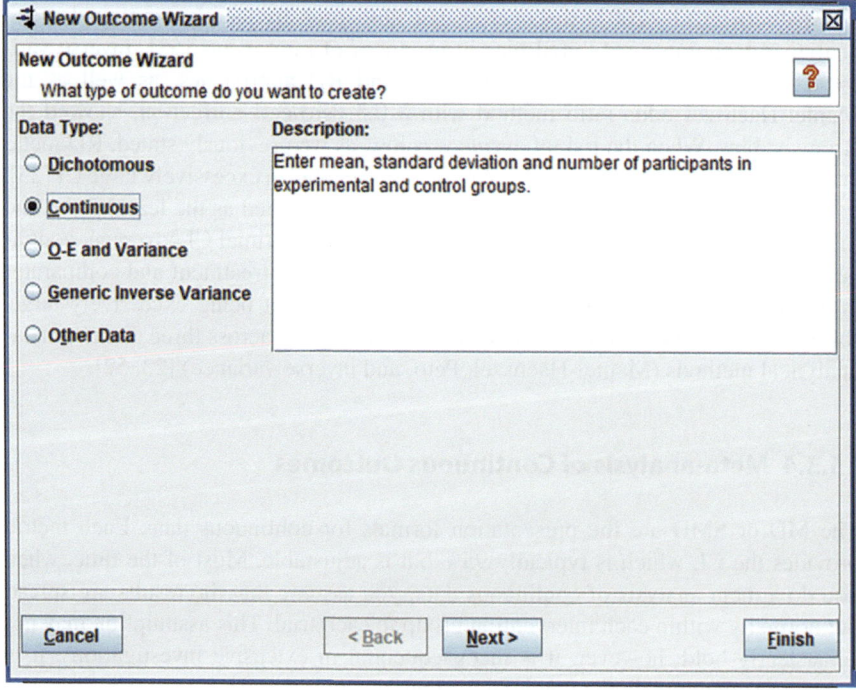

Fig. 15.11 Choosing the "continuous" method in RevMan

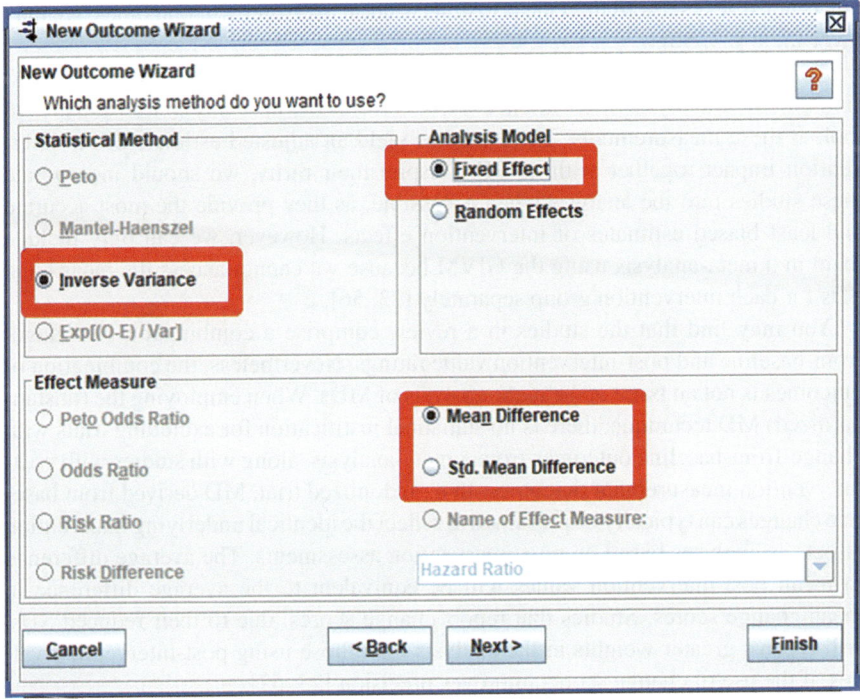

Fig. 15.12 Choosing analysis method according to the statistical method of continuous outcome in RevMan

assigned greater weight, whereas studies with bigger SDs receive lesser weight. This is fine if the differences in SDs between studies show that the outcome measures are not always accurate, but it probably is not right if the differences in SDs show real differences in how outcomes vary within the research populations [23].

The SMD method utilizes SDs to normalize MD to a uniform scale and to calculate studies' weights. Consequently, trials with small SDs yield comparatively greater estimates of SMD, whereas studies with larger SDs produce relatively lower estimates of SMD. To be deemed suitable, it must be presumed that the variation in SDs between studies alone represents discrepancies in measuring scales rather than differences in the reliability of outcome measures or variability among research populations [23, 55].

15.3.4.2 Meta-analysis of Change Scores

An analysis that uses changes from baseline may be more robust and effective than comparing post-intervention findings because it removes an element of inter-individual variability from the study. Calculating a change score necessitates measuring the outcome twice, which may be inefficient for unstable or challenging-to-measure outcomes where measurement error could exceed actual interindividual baseline variability. If the distribution of change-from-baseline outcomes is less skewed than the post-intervention measurement outcomes, we may favor them [23, 56].

Adding baseline measurements of the outcome variable as a covariate in a regression model or analysis of covariance (ANCOVA) is the best way to use statistics to look at these measurements. These studies yield an adjusted estimate of the intervention impact together with its SE. Despite their rarity, we should incorporate these studies into the analysis when accessible, as they provide the most accurate and least biased estimates of intervention effects. However, we can only include them in a meta-analysis using the GIVM because we cannot access the means and SDs for each intervention group separately [23, 56].

You may find that the studies in a review comprise a combination of change-from-baseline and post-intervention value ratings. Nevertheless, the combination of outcomes is not an issue in the meta-analysis of MDs. When employing the (unstandardized) MD technique, there is no statistical justification for excluding trials with change-from-baseline outcomes from a meta-analysis, along with studies with post-intervention measurement outcomes. In a randomized trial, MD derived from baseline changes can typically be presumed to reflect the identical underlying intervention effects as analyses based on post-intervention assessments. The average difference in mean post-intervention values will be equivalent to the average difference in mean change scores. Studies that report change scores, due to their reduced SDs, will receive greater weights in the analysis than those using post-intervention values, if the use of change scores improves precision [23, 57].

When authors put together data on the MD scale, they need to make sure they use the right means and SDs for each study. These can come from assessments done after the intervention or from changes compared to the baseline. Given that the mean values and SDs for the two outcome types may vary significantly, it is prudent to categorize them into distinct subgroups to prevent reader misunderstanding; nevertheless, the findings from these subgroups can be appropriately aggregated [23, 57].

Conversely, when the effect measure is an SMD, conventional meta-analysis methods should not, in theory, combine post-intervention values and change scores. This is due to the fact that the SD employed in the standardization represents distinct elements. The SD when standardizing post-intervention values indicates interindividual variability at a specific moment in time. The SD when standardizing change scores indicates the variety in interindividual changes over time, thus relying on both intraindividual and interindividual variability. Intraindividual variability is presumably influenced by the duration between measurements [23, 56]. Still, a real-world study of 21 osteoarthritis meta-analyses showed that there was no difference between SMDs that were calculated from values after the intervention and those that were calculated from change ratings [58].

15.3.4.3 Integration of Continuous and Dichotomous Outcomes

Certain research may represent data for the same outcome as dichotomous, while other research may present it as continuous. Researchers may present pain scale scores as means or as the percentage of patients experiencing pain at any point after an intervention, with a score exceeding a designated cutoff point. This information is typically more comprehensible and beneficial when dichotomized. Nevertheless, establishing a cut-point may be undefined, and information goes missing when continuous data is converted to dichotomous data [23, 59].

There are several ways to handle data combinations that are both continuous and dichotomous. Although it is generally desirable to compile results from all pertinent, reliable studies in a consistent way, this is not always possible. Obtaining missing data from investigators to aid in this procedure might be possible. Otherwise, there are three ways to summarize the data that can be useful:

- Presenting the data in textual format as "Other data" results
- Entering the counts as dichotomous outcomes
- Entering the means and SDs as continuous outcomes [23, 59]

There are statistical methods that can change ORs to SMDs and back again, which make it easier to combine discrete and continuous data [60]. This is a simple synopsis of the methodology: If the continuous measurements in each intervention group follow a logistic distribution, characterized by symmetry similar to the normal distribution but with more data in the tails, and if the outcome variability is the same for both experimental and comparator participants, we can reframe the OR as an SMD using the following formula:

- By multiplying by the constant ($\sqrt{3}/\pi = 0.5513$), the SE of the log OR can be converted into the SE of an SMD.
- As an alternative, SMDs can be multiplied by $\pi/\sqrt{3}$, or 1.814, to convert them to log OR.

You can use the GIVM to aggregate the SMD or log OR, along with their SE, for all studies in the meta-analysis. You can calculate the SE for all studies by inputting the data as dichotomous and continuous outcome types, where applicable, and transforming the CI for the resultant log OR and SMD into SE [61].

15.3.5 Measurement Scales and Ordinal Outcomes Meta-analysis

According to the approach used by the study authors in their initial analysis, ordinal and measurement scale results are frequently meta-analyzed as continuous or dichotomous data.

Data can occasionally be analyzed utilizing proportional odds models. Ordinal scales with only a few categories allow you to count the number of individuals in each category for each intervention group, and all trials use the same ordinal scale. This method may utilize all available data more efficiently than dichotomization; nevertheless, it necessitates access to statistical software and produces a summary statistic that is difficult to interpret clinically [23, 62].

The proportional odds model measures the impact of an intervention using the proportional OR [63]. More advanced statistical software (R Software, OpenMeta[Analyst], or Comprehensive Meta-Analysis) can also use it for meta-analysis. The GIVM can be used to do meta-analysis on the log OR and the SE that come with it when we use a proportional odds model. Let us assume that all trials use a consistent ordinal scale, despite some publications classifying it as a dichotomous outcome. In that case, it remains feasible to incorporate all studies in the meta-analysis [64].

15.3.6 Meta-analysis of Counts and Rates

When each person experiences an event more than once, the results can be shown as count data. For example, "number of hospital visits" and "number of strokes" are measurable statistics. Theoretically, a person can have as many of these events as they like, but they might never occur. In addition to being evaluated as rate data, count data can also be studied using techniques for time-to-event data, continuous data, and dichotomous data if the counts are dichotomized for each individual [23, 65].

Recording counts for each participant along with their observation duration yields rate data. This becomes particularly relevant when the enumerated events are rare. For example, over a two-year follow-up period, a woman might experience two strokes. According to follow-up, she had one stroke every year, or 0.083 every month. Group-level rates frequently combine.

- **Example**: Over a total of 2836 person-years of follow-up, participants in a clinical trial's comparison group might experience 85 strokes. When using rates, it is implicitly assumed that the risk of an occurrence stays constant across time and among individuals. Every situation requires a careful assessment of this premise. Analyzing count data as rates is rarely, if ever, improper. This is because the statistical techniques are less sophisticated than those for other data kinds, and it is assumed that a stable underlying risk may not be acceptable and the statistical methods are less advanced compared to those for other data types [23, 65].

The results of a study can be shown using a rate ratio, which compares the rates in the experimental and control groups. The GIVM can be used to aggregate the rate ratios' natural logarithms across trials. Alternatively, Poisson regression methods may be conducted [54, 66].

Being expecting similar average follow-up times for each intervention group, rate ratios in a randomized trial often resemble risk ratios derived from dichotomizing participants. If an intervention affects a participant's likelihood of experiencing several occurrences, rate ratios and risk ratios will differ [23, 66].

15.3.7 Meta-analysis of Time-to-Event Outcomes

Time-to-event outcomes, also known as survival outcomes or failure-time outcomes, refer to the analysis of the time it takes for a specific event or outcome of interest to occur. These events could include things like disease progression, death, relapse, recovery, or any other event that happens after a defined starting point [13].

Time-to-event outcomes are frequently utilized in the medical field. These outcomes are concerned not only with the occurrence of the event (i.e., dichotomous data) but also with the time frame that has passed from a clearly defined starting point. Medical statistics literature commonly refers to them as survival data, given that death often serves as the primary event of interest.

The data of time to event of each participant consists of two observations:

– First, a time period during which nothing happened
– Then, a signal that says whether an event or merely the end of observation occurs when that time frame ends
– **"Censored"** is used to describe participants who give time that does not result in an event. The analysis incorporates the data from their event-free duration.

Time-to-event statistics can depend on events rather than mortality, such as disease recurrence or hospital discharge [13]. This necessitates knowing the condition of each experiment participant at a given point in time to examine time-to-event data as a binary outcome. If all the patients have been tracked for a certain duration and the percentage of patients in each group who had the event before this duration is known, the intervention's effects can be analyzed as RR, OR, HR, rate ratio, or RD (Table 15.3).

Since pertinent times are only accessible for the subset of people who have experienced the event, it is improper to analyze the time-to-event data using techniques intended for continuous outcomes. Additionally, the exclusion of censored participants is necessary, which will likely cause bias [23].

The optimal strategy for summarizing time-to-event data is through survival analysis, articulating the intervention effect as a HR. Hazard is analogous to risk; however, it differs considerably in that it quantifies immediate risk and can fluctuate continuously. A HR tells you how likely it is that someone will experience something at a certain time while getting the experimental intervention compared to the comparison intervention. Meta-analyses and therapy trials often assume the HR stays the same during follow-up, even though risks can change. This is referred to as the proportional hazard assumption [13].

The meta-analysis of time-to-event results indicates that pooled estimates can be determined using both random-effect and fixed-effect analyses, which are easily accomplished with statistical software (e.g., RevMan employing GIVM or Stata utilizing the "metan" command [Stata Corp., College Station, Tex]) [67].

Therefore, authors have easy access to two available approaches for meta-analyzing time-to-event results. Which one to use will depend on the type of data obtained from the original study or obtained by reanalyzing data from specific participants.

"O–E" and "V" statistics can be immediately entered into RevMan using the "O–E and Variance" outcome type if they have been obtained, either by re-analysis of individual participant data or from aggregate statistics described in the study reports. Multiple methods exist for calculating the "O–E" and "V" statistics. Peto's method for dichotomous data yields an OR; a log-rank method produces a HR; and a modified Peto method for time-to-event analysis results in an intermediate measure [68]. We use only the fixed-effect method of meta-analysis for "O–E and Variance" outcomes.

Alternatively, GIVM can be used to aggregate trials' findings if estimates of log HR and SE have been obtained via Cox proportional hazards regression models. Since log-rank estimates may be converted into log HR and SE, the GIVM can be used to integrate all results when a mix of log-rank and Cox model estimates are available from the study data.

A time-to-event outcome is entered into the RevMan application in two ways:

- It can be entered by O–E and variance.
- All steps are the same as the previous steps except steps 4 and 6 (Table 15.4).

Step 4: In the New Outcome Wizard window, choose "O–E and variance." (Fig. 15.13).

Step 6: The statistical method is "Exp[(O-E)/Var]"; the analysis method will be a "fixed-effect" model, and the effect measure will be the "Peto Odds Ratio" (Fig. 15.14).

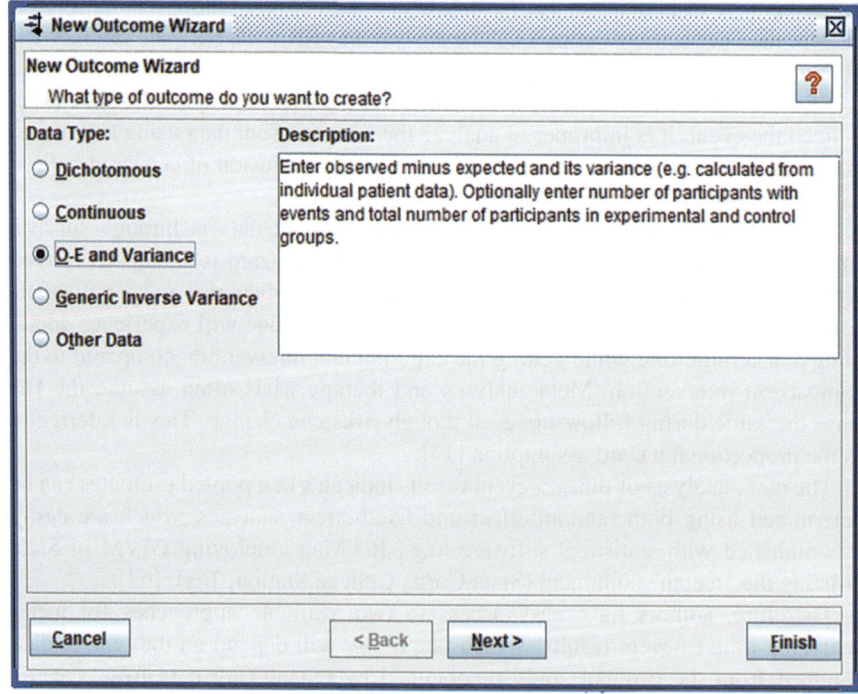

Fig. 15.13 Choosing the "O–E and variance" statistical method in RevMan

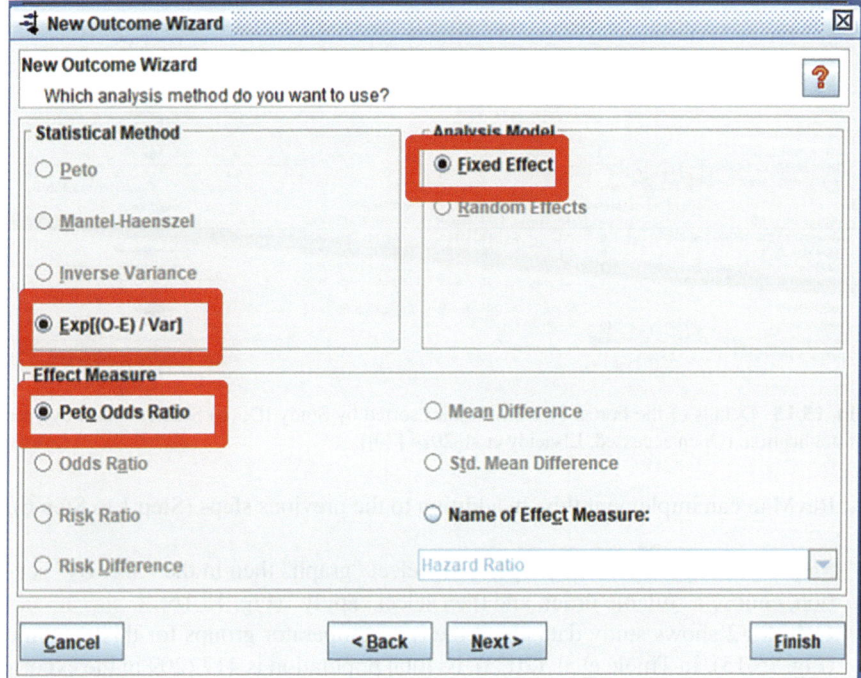

Fig. 15.14 Choosing the "Exp[(O-E)/Var]" statistical method in RevMan

Which effect measure for the continuous outcomes will be discussed in the next section.

- It can be entered by GIVM. All steps of GIVM were mentioned before.

15.4 Displaying Results

15.4.1 Details of the Forest Plot

The forest plot is considered a visualization tool for meta-analysis, which facilitates understanding the relationship between the two groups (intervention and control groups) [69]. It consists of six columns, as presented in Fig. 15.15a [70].

- **Column 1** labeled "Study or Subgroup," indicates the included studies in this meta-analysis, labeled by the last name of the first author + et al. + year of publication, as in Thiele et al. (2023). The list of studies can be sorted by study ID (Fig. 15.15a) or by year of publication (Fig. 15.15b).
  Column 1. (Study Details).; **Column 2.** (two comparative groups).; **Column 3.** (weight).; **Column 4.** (study effect measure).; **Yellow arrow.** Low-influenced study.; **Red arrow.** High-influenced Study.; **Green arrow.** Overall effect.; **Blue arrow.** Line of no effect.

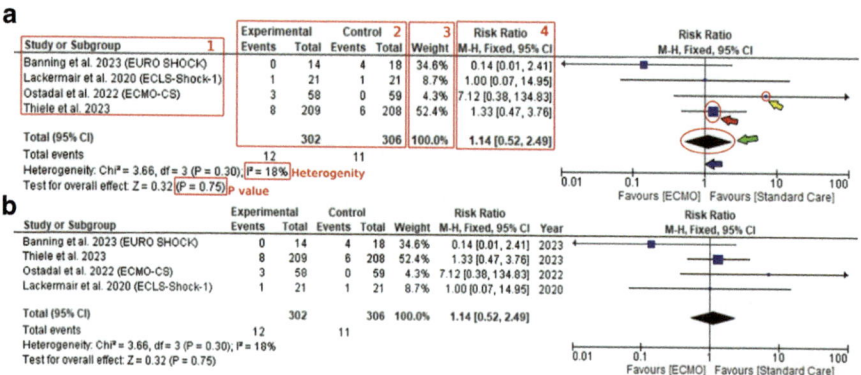

Fig. 15.15 Details of the Forest Plot: (**a**) Studies sorted by Study ID; (**b**) Studies sorted by year of publication. (Open accessed, Elsaeidy et al. 2024 [70])

RevMan can implement this. In addition to the previous steps (Step 1 to Step 6),

- **Step 7:** through "outcome properties," select "graph" then in the "Sort By" section, choose a suitable order, and then select "apply" (Fig. 15.16).
- **Column 2** shows study data about the two comparator groups for this outcome (Fig. 15.15). In Thiele et al. (2023), its total population is 417 (209 in the experimental cohort and 208 in the control cohort).
- **Column 3** indicates the influence of each study in the overall meta-analysis, as Thiele et al. (2023) is the most influencing study with a weight of (58.2%), and Ostadal et al. (2022) is the lowest with a weight of (13.0%). A large weight is for studies that will report more information, usually studies with a large sample size [71].
- **Column 4** reports the numerical results of each study and the overall meta-analysis results (for example, OR, RR, and 95% *CI* for dichotomous outcomes, MD, or SMD, and 95% *CI* for continuous outcomes) such as Banning et al. (2023) (0.14 [0.01,2.41]) and overall results (1.16 [0.38,3.57]) (Fig. 15.15).

A wide *CI* (the longer the line in the forest plot) indicates less precise study data [72]. The middle vertical line in the graph is the line of no effects; its value is often one for dichotomous outcomes and zero for continuous outcomes, indicating that there is no statistically significant difference between the two comparator groups [71]. Studies with CI that do not overlap the line of no effect indicate that it will report statistically significant data like Thiele et al. (2023) [73] in Fig. 15.15 [69, 74]. The diamond shape in the graph indicates the overall meta-analysis results; if it does not cross the vertical line of no effect, this indicates a significant difference between the two groups; if it crosses the vertical line, this indicates an insignificant difference between the two groups, preferably the one to which it belonged

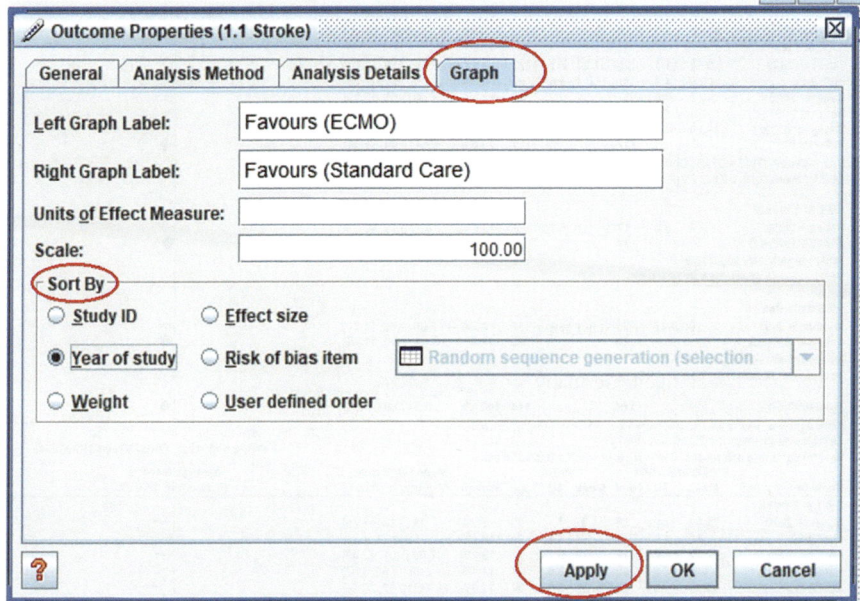

Fig. 15.16 Study order in RevMan

(experimental group or control group). The overall significance of the data is detected by the *P value*, as the result is considered statistically significant if $P < 0.05$ [69, 71].

15.4.2 Group Labeling in Review Manager Software (RevMan)

Group labelling is a very important step in meta-analysis as it may lead to misinterpretation. RevMan's assumption assumes that all studies discuss unfavorable outcomes and label intervention A on the left axis and control on the right axis, even if it is incorrect according to the pooled effect estimate and CI. Therefore, it depends on the statistician to label each group correctly to be suitable for the pooled effect estimate and CI. Elsaeidy et al. (2024) assessed the efficacy and safety of levosimendan in patients with advanced heart failure [75]. The outcome of interest was left ventricular ejection fraction (LVEF) (Fig. 15.17a). The authors edited the group label to be suitable for the clinically favorable outcome and put the intervention arm on the right of the X-axis of the forest plot, which declares that levosimoendan is associated with an enhancement in LVEF; otherwise, if the author keeps it according to the RevMan assumption, it will be misinterpreted and inaccurate in the meta-analysis (Fig. 15.17b). Therefore, it is essential to understand the outcomes and compare the visual data of the forest plot (diamond shape) with the pooled effect estimates and CI.

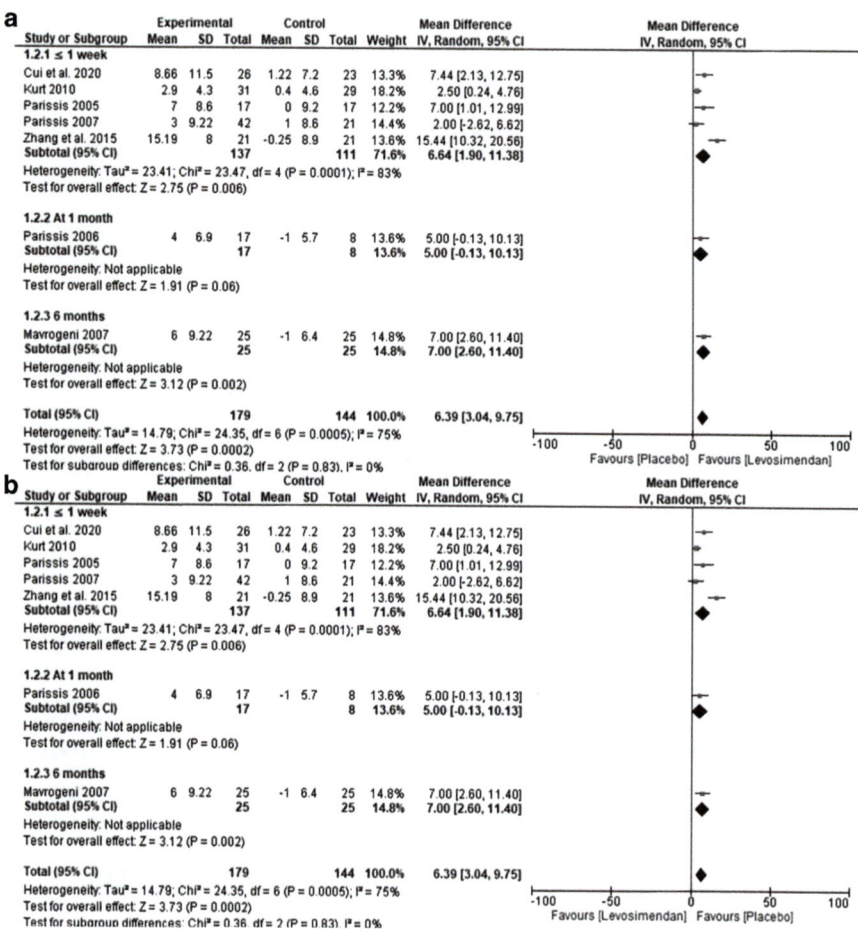

Fig. 15.17 A guide to graph labeling. (**a**) Authors assigned the intervention group to the right graph; (**b**) default assignment of the intervention group in RevMan is to the left graph. (Open accessed, Elsaeidy et al. [75])

15.5 Heterogeneity

15.5.1 What Is Heterogeneity?

Heterogeneity is a fundamental concept in meta-analysis and refers to the differences between studies that arise from different populations, study designs, strategies, outcome measures, and time points. Heterogeneity can be reduced by establishing comprehensive PICO criteria or using a suitable randomization tool or allocation concealment, but it cannot be eliminated [11, 23, 76].

15.5.2 Identification and Measuring Heterogeneity

There are three types of heterogeneity in meta-analysis:

- **Clinical heterogeneity** refers to variations in study participants, interventions, and other demographic data [23].
- **Methodological heterogeneity** refers to variations in study design and outcome measurement tools [23].
- **Statistical heterogeneity** is present when there is a variations in the true effect between the studies. This type of heterogeneity can be assessed through some statistical tests, such as the I-squared (I^2) statistic and Cochran's Q test [77].

15.5.2.1 Cochran's Q Test
Cochran's Q test is considered as the sum of the squared deviations of the study effects from the overall effect with each study contribution. It is essential to note that the Q test is not by itself a measure of heterogeneity. Instead, it is used to test the hypothesis that there is no heterogeneity, and P values are reported. When the P value <0.10 suggests the presence of heterogeneity. Although the Q test is widely used, it may not be suitable for detecting heterogeneity in meta-analyses with few studies, as the values of Q and the P value of the Q-statistics depend on the number of included studies in the analysis [71, 76, 78].

15.5.2.2 The I-Squared (I^2) Statistic
The second commonly used test to detect heterogeneity is the I-squared (I^2) statistic, which represents the percentage of variability in the effect estimate across all studies attributing heterogeneity [71]. The I^2 statistic varied from 0 to 100%. Typically, if the I^2 value ranges from 0 to 40%, it might be considered unimportant; 30–60% might be moderate heterogeneity; 50–90% might be substantial heterogeneity; and 75–100% might be considerable heterogeneity [23].

I^2 is commonly presented as a single value without a 95% CI; therefore, it provides a summary of existing heterogeneity that impacts the meta-analysis [71].

Figure 15.18 shows that the included studies in the meta-analysis are homogeneous, as the P value of Cochran's Q test was 0.30 and I^2 was 18%.

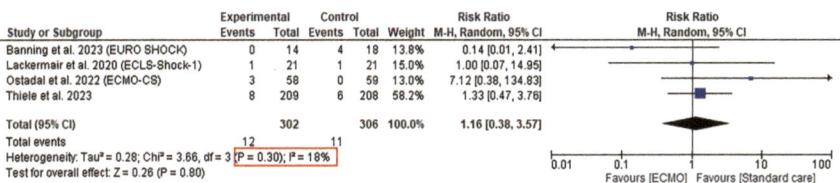

Fig. 15.18 Meta-analysis of homogeneous studies. (Open accessed, Elsaeidy et al. [70])

15.5.3 Addressing Heterogeneity and Sensitivity Analysis Test

15.5.3.1 Incorporating Heterogeneity into Random-Effects Models

In meta-analysis, there are two types of statistical analysis models: fixed and random. The fixed model assumes that variations between studies are attributable to sampling errors or chances [27]. This model only assessed intra-study variation, which is statistically applicable when there is no significant heterogeneity among studies (heterogeneity $P \geq 0.10$) or when the I^2 statistic is below 50% ($I^2 < 50\%$) [27]. For example, Shaheen et al. (2024) used a fixed model as there was no heterogeneity ($I^2 = 0\%$) in the occurrence of seizures outcome (Fig. 15.19) [79].

The random-effects model, on the other hand, evaluates whether the difference between studies is attributable to variations in their true effect sizes, which evaluates both intra-study sampling errors and inter-study variance (between-study variation) [27]. This variation can be statistically detected when significant heterogeneity is present (heterogeneity $P < 0.10$) or when the I^2 statistic exceeds 50% ($I^2 > 50\%$) [27]. For example, Elsaeidy et al. (2024) used the random model as there was substantial heterogeneity ($I^2 = 70\%$, $P = 0.04$) in all-cause rehospitalization outcome (Fig. 15.20a) [75].

We can edit the fixed effects to random effects by pressing on them, as shown in Fig. 15.21.

15.5.3.2 Subgroups

Subgroup analysis in meta-analysis is a tool to split the included studies based to patient characteristics such as age, sex, race, and other baseline characteristics, or study characteristics such as intervention type and outcome measures [80]. We can

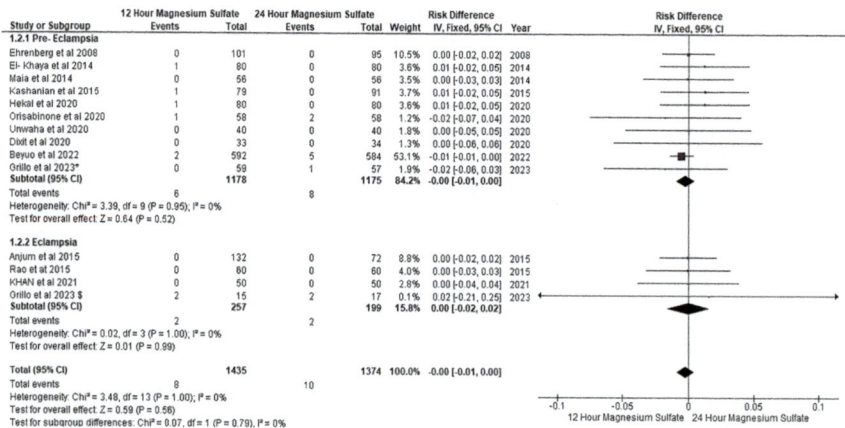

Fig. 15.19 Forest plot of fixed model subgroup meta-analysis according to patient category (disease). (Open accessed, Shaheen et al. [79])

a

Study or Subgroup	Experimental Events	Total	Control Events	Total	Weight	Risk Ratio M-H, Random, 95% CI	Risk Ratio M-H, Random, 95% CI
Comin-Colet et al. 2018	17	48	15	21	33.4%	0.50 [0.31, 0.79]	
Cui et al. 2020	0	26	0	23		Not estimable	
Garcia-González et al. 2021	23	70	12	27	30.4%	0.74 [0.43, 1.27]	
Llorens et al. 2012	18	25	13	20	36.2%	1.11 [0.74, 1.66]	
Total (95% CI)		169		91	100.0%	0.75 [0.46, 1.22]	
Total events	58		40				

Heterogeneity: Tau² = 0.13; Chi² = 6.62, df = 2 (P = 0.04); I² = 70%
Test for overall effect: Z = 1.15 (P = 0.25)

0.01 0.1 1 10 100
Favours [Levosimendan]   Favours [control]

b Table S5. Leave one out sensitivity analysis of all-cause rehospitalization rate at 6 months.

Study	All-cause rehospitalization at 6 months						
	Effect size				Heterogeneity		
	Risk ratio	Lower bound	Upper bound	P-value	I-squared	P-value	
Overall	0.715	0.424	1.204	0.207	72%	0.03*	
Comin-Colet et al. 2018	0.876	0.509	1.509	0.634	63%	0.1	
Gracia-Conzalez et al. 2021	0.748	0.34	1.644	0.47	85%	0.01*	
Llorens et al. 2012	0.543	0.375	0.786	0.001*	0%	0.53	

Fig. 15.20 Forest plot of random model meta-analysis. (**a**) Meta-analysis with high heterogeneity without sensitivity analysis; (**b**) meta-analysis after sensitivity analysis (leave-one-out test). (Open accessed, Elsaeidy et al. [75])

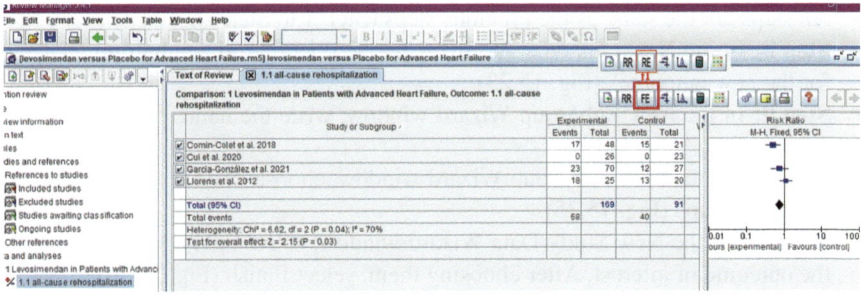

Fig. 15.21 Changing fixed to random model in RevMan. Random-effect model (RE); fixed-effect model (FE)

also conduct a subgroup analysis to identify the heterogeneity, as it provides more information about interventions' effect on each group and the variability across the included studies 23. Therefore, it may be in a fixed model or random model according to the heterogeneity in each subgroup; if one subgroup only has significant heterogeneity, a random model should be used [23]. For example, Shaheen et al. [79] conducted subgroup analysis according to the patient category, as in the occurrence of seizure outcome, and there was no heterogeneity between 12-h and 24-h magnesium sulfate in any subgroup; thus, they used a fixed model (Fig. 15.19). On the other hand, Elsaeidy et al. (2024) performed a subgroup analysis based on patient age. In the tachycardia outcome, there was substantial heterogeneity

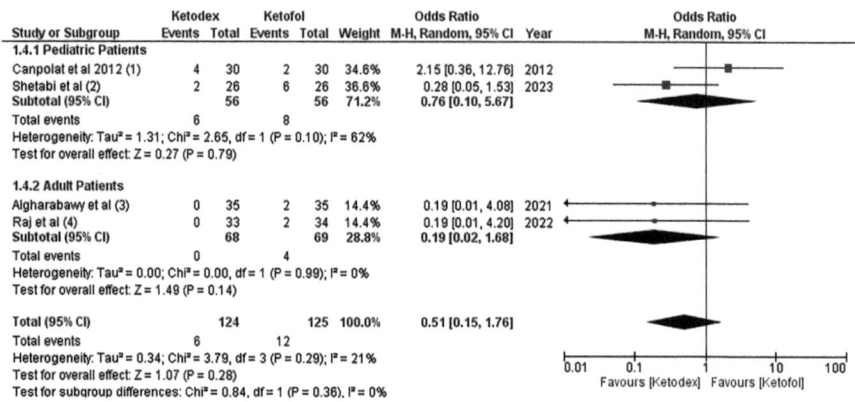

Fig. 15.22 Subgroup meta-analysis according to baseline characteristics (patients age). (Open accessed, Elsaeidy et al. [81]

($I^2 = 62\%$) in the pediatric patient subgroup; therefore, they used a random model (Fig. 15.22) [81].

- Subgroup analysis can be conducted in RevMan through some steps.
- All steps are the same as the previous steps (Step 1 to Step 7).
- **Step 8:** In the New Outcome Wizard window, select "Add a subgroup analysis for the new outcome" (Fig. 15.23).
- **Step 9:** In the New Subgroup Wizard window, write the name of the subgroup (Fig. 15.24).
- **Step 10:** In the New Subgroup Wizard window, choose "Add study data for the new subgroup" (Fig. 15.25).
- **Step 11:** In the New Study Data Wizard window, select the included studies in the outcome of interest. After choosing them, select Finish (Fig. 15.26).
- There is another method in the intervention review page under the "Data and analyses" tab: select a comparison, then choose an outcome. Right-click here, and you'll find many demands click "Introduce subgroup" (Fig. 15.27).

15.5.3.3 Meta-regression

Meta-regression is considered an extension of subgroup analysis, as subgroup analysis assesses the relationship between study characteristics and intervention effect, but meta-regression assesses the association between the effect size and multiple factors of study characteristics, including those not included in subgroup analysis, which gives a comprehensive understanding of factors that affect the effect size [80]. Like simple regression meta-regression had an outcome variable that predicted

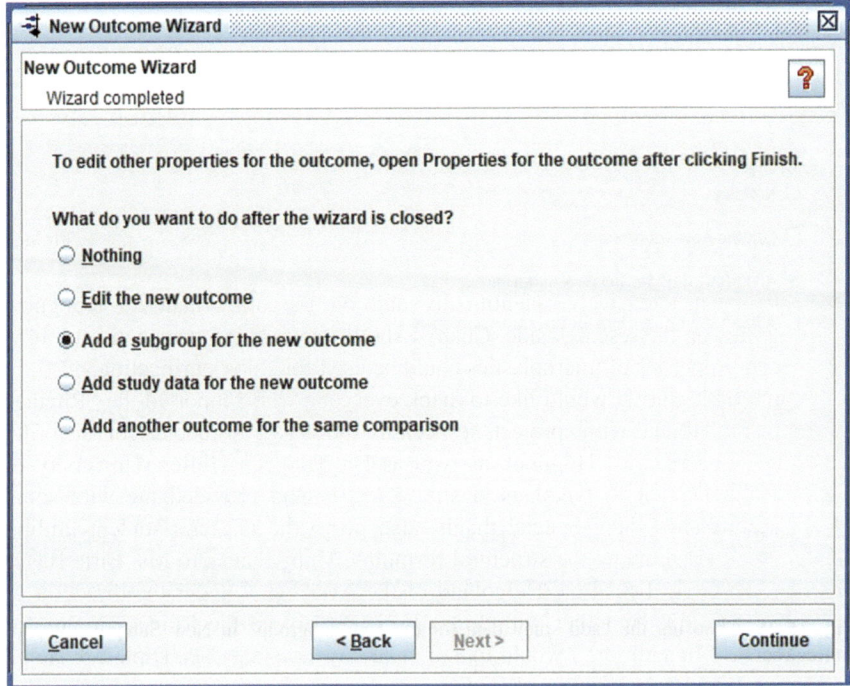

Fig. 15.23 Choosing the "add a subgroup for the new outcome" in New Outcome Wizard in RevMan

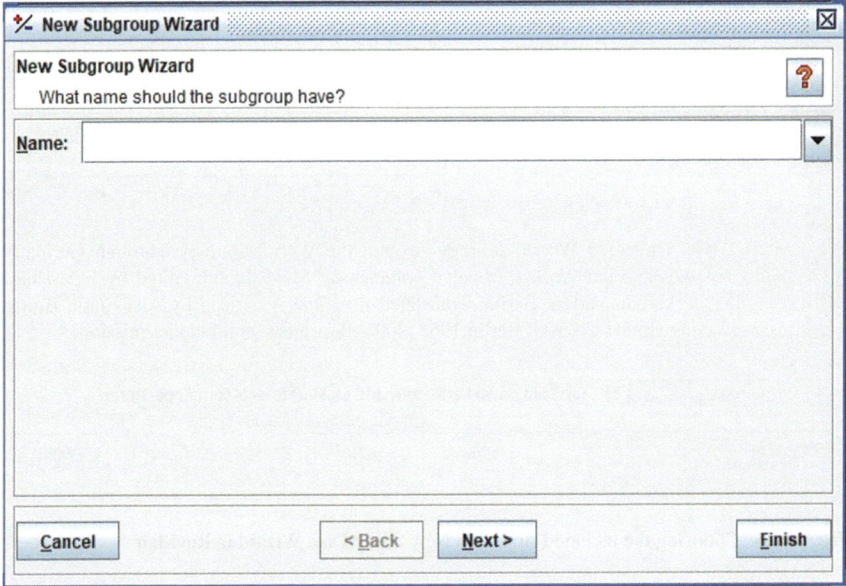

Fig. 15.24 Name the subgroup in New Subgroup Wizard in RevMan

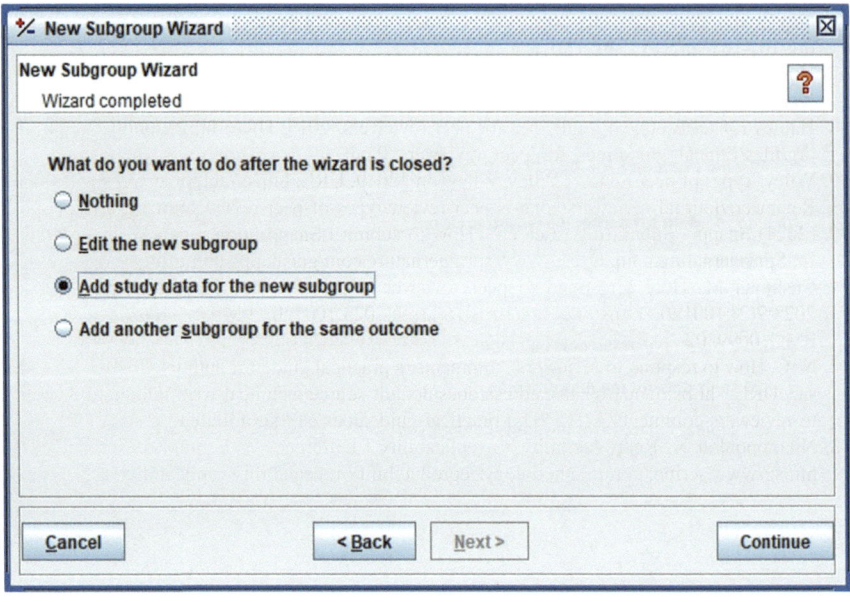

Fig. 15.25 Choosing the "add study data for the new subgroup" in New Subgroup Wizard in RevMan

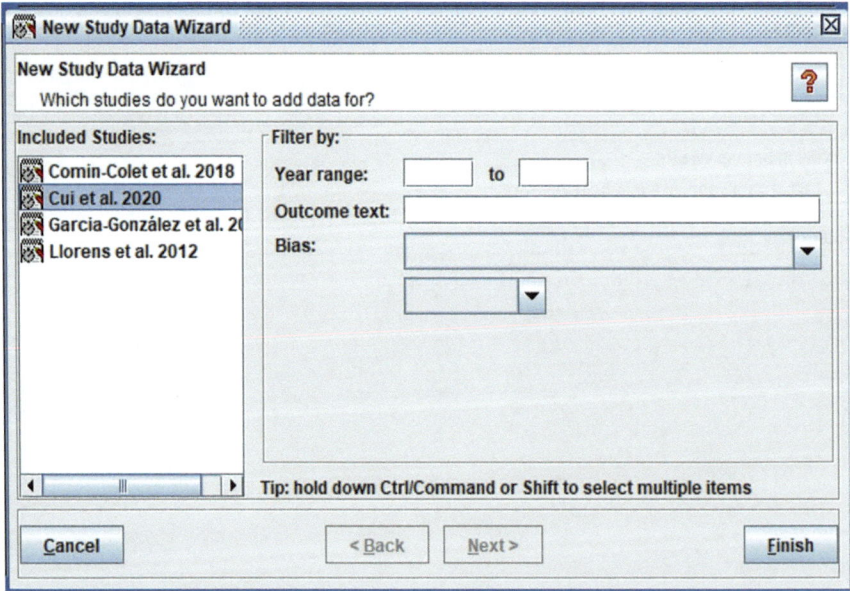

Fig. 15.26 Choosing the included studies in New Study Data Wizard in RevMan

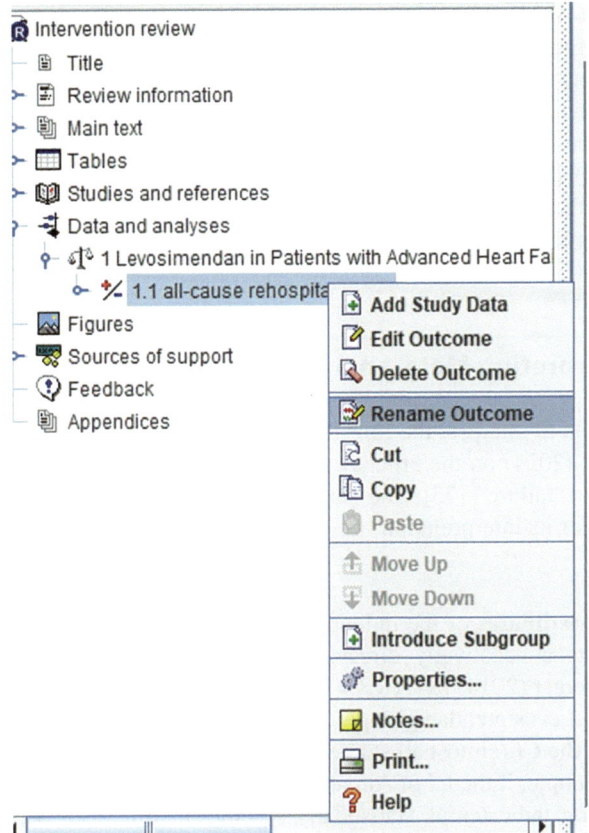

Fig. 15.27 Choosing "Introduce Subgroup" in RevMan

by one or more explanatory variables. In meta-regression, the outcome variable is considered the effect size, which may be a MD, risk difference, log OR, or log RR according to the outcome of interest, and the explanatory variables are characteristics of studies that might influence the intervention's effect size. [27] Meta-regression can be conducted when **ten or more** included studies report the outcome of interest. [27]

15.5.3.4 Leave-One-Out Test

The leave-one-out test is the most common test used to assess the sensitivity of the results. [82] We performed this test if there was significant heterogeneity $I^2 > 50\%$ by excluding each study on the individual case and recalculating the results. [83] If there is a specific study that makes a significant difference and reduces heterogeneity, it indicates that this study significantly affects the overall results. [83] We can do this by using

- **OpenMeta-Analyst** (http://www.cebm.brown.edu/openmeta/).
- **STATA** (https://download.stata.com/).
- **RevMan**, but it does not report a specific figure for the leave-one-out test; therefore, we can report two figures of the outcome, one before and the other after the leave-one-out test, or report the leave-one-out test in a table, such as Elsaeidy et al. (2024). [75] The outcome of interest is an all-cause rehospitalization rate at 6 months; there was considerable heterogeneity (I^2 = 70%) (Fig. 15.20a), and after the leave-one-out test by excluding Llorens et al. (2012), there was no heterogeneity (I^2 = 0%) (Fig. 15.20b).

15.6 Interpreting Meta-analysis

For guiding you to interpret the forest plot, we used a meta-analysis conducted by Elsaeidy et al. (2024) on the efficacy and safety of levosimendan in patients with advanced heart failure. [75] The outcome of interest was all-cause mortality (Fig. 15.28). For its interpretation, we should consider some key points as reported in Table 15.5.

- **The effect estimates** of the individual included study, in Fig. 15.28 almost all studies have a negativelly directed effect estimate (RR < 1.0). For example, Altenberger (2014) [84] (RR = 0.23) declares that the risk of all-cause mortality in the Levosimendan group was less than the risk in the placebo group.
- We look at **the *CI*** *(almost all studies have a narrow CI)*, which indicates precise data for example, if the CI of Altenberger et al. (2014) = (0.10, 0.50).
- **The *P* value** indicates as statistically significant difference between the two interventions if it was < 0.05 (as in the all-cause mortality (Fig. 15.28), the *P value* = 0.01, which indicates that levosimendan is associated with a significant decrease in the incidence of all-cause mortality compared with placebo).
- **The heterogeneity *Q* test *p* value,** *which* was 0.36 (in Fig. 15.28) indicating that the variability across the studies was not statistically significant. At the same time, I^2 statistic was 9%, indicating unimportant heterogeneity.
- The last item that should be taken into consideration is **the overall effect.** The overall effect was RR 0.60 [95% *CI* 0.40–90]), indicating that levosimendan is beneficial in reducing the incidence of this outcome.

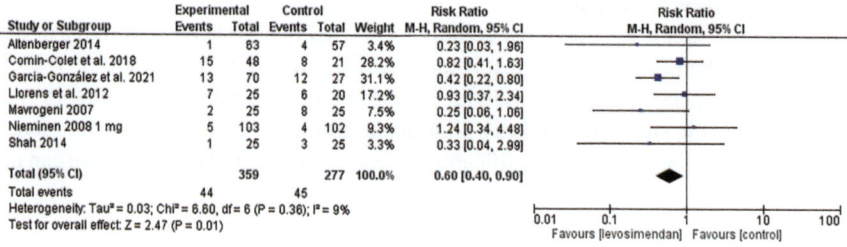

Fig. 15.28 How to interpret a forest plot. (Open accessed, Elsaeidy et al. 2024 [75])

Table 15.5 Key points to be considered in the interpretation of the meta-analysis

Key points	Description	How to interpret
Effect estimate	The measured effect of each study (e.g., odds ratio, relative risk, mean difference)	Look for the direction of the effect (positive or negative)
Confidence interval (CI)	Range of values for each study	Narrow CI indicates greater precision
P value	The probability of the observed effect	Often used to determine statistical significance, small p values (<0.05) suggest statistically significant results
Heterogeneity	Variability between the included studies	Assessed using statistical tests (e.g., Q test, I^2 statistic)
Overall effect	Summary of the overall effect across all studies	Indicates the overall direction of the effect

Finally, we can interpret the outcome as follows:

Seven studies with 636 patients were included. We found that levosimendan was associated with a decreased incidence of all-cause mortality compared with placebo (RR: 0.60, 95%CI [0.40, 0.90], P = 0.0.01), with no statistically significant heterogeneity (I^2 = 9%, P = 0.36).

15.7 Problems and Limitations in Conducting Meta-analysis

Inappropriate meta-analyses can produce misleading outcomes, either false-negatives or false-positives. False-positives, in particular, can postpone the start of large, conclusive trials, as some meta-analyses of smaller studies have been later disproven by extensive randomized controlled trials. Various factors contribute to the unreliability of meta-analysis results, highlighting the importance of evaluating the quality of the included studies, the variability of the results, and the presence of metabias when critically assessing a meta-analysis [6].

15.7.1 Poor Quality of the Included Trials

One notable limitation is described as "garbage in, garbage out." This implies that if a meta-analysis incorporates poor-quality, biased trials, the resulting conclusions will also be biased and inaccurate. Consequently, the dependability of the results largely depends on the quality of the trials it encompasses [85]. Key quality elements of randomized trials include the concealment of treatment allocation procedures, blinding of participants and personnel, minimal and well-documented losses to follow-up, and intention-to-treat analysis. So, it is crucial to evaluate the methods used in the included studies in a standardized approach.

Solution
Sensitivity analyses can evaluate the reliability of the meta-analysis results by comparing pooled outcomes from high-quality and poor-quality trials. Meta-analysis results are more reliable when confirmed by separate analyses of high-quality studies [23]. Meta-analyses that incorporate observational studies should be handled carefully, as they may produce highly precise but potentially misleading results as a result of confounding factors and selection bias [86].

15.7.2 Heterogeneity in Study Outcomes

Another significant criticism of meta-analysis is its tendency to incorporate various types of studies, similar to "mixing apples and oranges." While meta-analysis objectively and quantitatively assesses variations and inconsistencies (heterogeneity) in study results (providing a more unbiased estimate on a given topic), it should be avoided if the studies are excessively heterogeneous, making them incomparable [87].

Meta-analysis of controlled trials assumes that each included study offers an unbiased estimate of the intervention's effect, with any variability being due to random variation.

Solution
- To address this issue, one can evaluate the variability among the studies and conduct subgroup analyses [87].
- In case of the substantial variation in point estimates, non-overlapping 95% CI, and a large I^2 in a forest plot which indicates significant heterogeneity, determining a combined effect size might be unsuitable. Therefore, the sources of heterogeneity should be explored through meta-regression or subgroup analysis (see 5.3.2 and 5.3.3) [88].

15.7.3 Metabias

- **Publication bias**, a type of metabias, is a significant concern in research. Publication bias can result in an overestimation of treatment effects, as studies showing positive outcomes are more likely to be published compared to those showing negative or inconclusive findings [89, 90]. For instance, a study by Turner et al. found that while 94% of trials on antidepressants appeared positive in published literature, only 51% were positive according to FDA reviews. This discrepancy highlights how publication bias can distort the perceived efficacy of treatments [91]. To minimize publication bias, comprehensive systematic

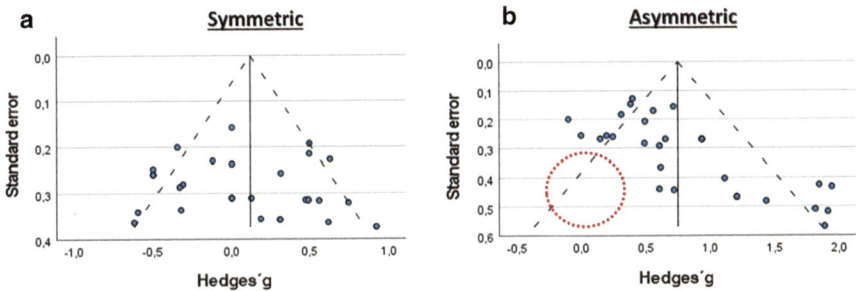

Fig. 15.29 Visual inspection of a funnel plot

literature searches are essential. This involves searching multiple electronic databases, conference proceedings, trial registries, and reaching out to experts and pharmaceutical companies. The PRISMA statement also emphasizes the importance of transparency in reporting all information sources and search strategies used in meta-analyses [7, 8].
– **Reporting bias** arises when the publication of research outcomes is affected by the nature and direction of the results. Detecting reporting bias, especially when some studies or outcomes remain unpublished, can be challenging. One of the widely used methods is the funnel plot (Fig. 15.29).

The illustration depicts a visual inspection of a **funnel plot** to assess the risk of publication bias. **Blue dots** represent individual studies. A symmetrical funnel plot (**a**) is usually interpreted as reflecting an absence of publication bias, with studies evenly distributed around the central axis. In contrast, an asymmetrical funnel plot (**b**) typically indicates the presence of publication bias, as studies cluster unevenly. The **red dotted circle** denotes an empty space where studies were expected to be present but are missing. (Open accessed, Afonso et al. 2023 [96]).

A funnel plot is a scatter plot done to assess reporting bias in meta-analyses [92]. It plots the effect estimates from individual studies against a metric of each study's size or accuracy, often using the SE of the effect estimate. Studies with greater precision and larger sample sizes are positioned higher up, while those with less precision and smaller sample sizes are more widely dispersed at the bottom. If there is no bias, the plot is expected to create a symmetrical, inverted funnel shape. Asymmetry in the shape of the funnel plot may refer to reporting bias, but it can also result from other factors like methodological limitations in smaller studies or chance [93].

- **N.B.** A visual inspection of a funnel plot is used to evaluate the risk of publication bias (Fig. 15.29). With blue dots representing individual studies. Symmetry in the plot suggests no bias, while asymmetry indicates potential bias. The red dotted outline highlights missing studies. When a funnel plot shows asymmetry, statistical methods are employed to understand how potential publication bias

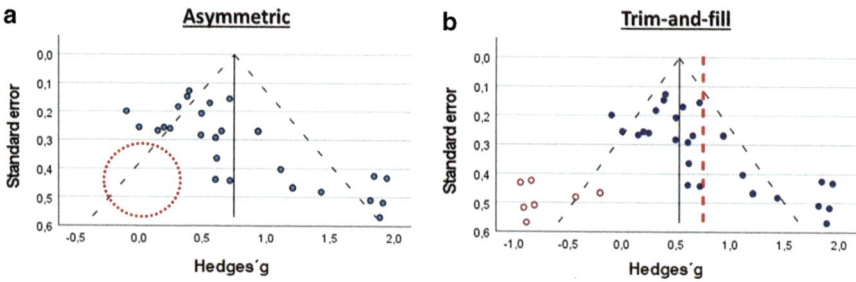

Fig. 15.30 Trim-and-fill statistical test

might affect the overall effect size. For instance, the Trim-and-fill method adjusts for publication bias by iteratively removing extreme small studies from the positive side of a funnel plot, and then imputing missing studies to balance it (Fig. 15.30). This generates a less biased summary effect size but relies on the assumption that such missing studies exist. It should be used primarily for sensitivity analyses.

The illustration demonstrates the **trim-and-fill** statistical test used to correct funnel plots for publication bias. **Blue dots** indicate individual studies. (**a**) Absence of studies on the left side of the funnel plot is highlighted by a **red dotted circle**. (**b**) Trim-and-fill correction, where imputed studies (**red circles**) are added, and the pooled Hedge's g is adjusted. The previous pooled effect size is marked by a red dotted line. (Open accessed, Afonso et al. [96])

- **Early stopping for benefit bias** is another metabias where studies can overestimate treatment effects, especially if they have a small number of events or substantial weight in the meta-analysis [94].
- Another metabias is that **single-center RCTs may overestimate treatment effects** compared to multicenter trials on the same topic, resulting in inaccuracy in summary estimates in meta-analyses [95].

References

1. Murad MH, et al. How to read a systematic review and meta-analysis and apply the results to patient care: users' guides to the medical literature. JAMA. 2014;312:171–9.
2. Guyatt GH, et al. Users' guides to the medical literature: IX. A method for grading health care recommendations. JAMA. 1995;274:1800–4.
3. Egger M, Smith GD, Phillips AN. Meta-analysis: principles and procedures. BMJ. 1997;315:1533–7.
4. Gøtzsche PC. Why we need a broad perspective on meta-analysis: it may be crucially important for patients. BMJ. 2000;321:585–6.
5. Walker E, Hernandez AV, Kattan MW. Meta-analysis: its strengths and limitations. Cleve Clin J Med. 2008;75:431–9.

6. Meta-analyses: what they can and cannot do | Swiss Medical Weekly. https://smw.ch/index.php/smw/article/view/1434.
7. Moher D, Liberati A, Tetzlaff J, Altman DG, Group TP. Preferred reporting items for systematic reviews and meta-analyses: the PRISMA statement. PLoS Med. 2009;6:e1000097.
8. The PRISMA statement for reporting systematic reviews and meta-analyses of studies that evaluate health care interventions: explanation and elaboration | PLOS Medicine. https://journals.plos.org/plosmedicine/article?id=10.1371/journal.pmed.1000100
9. Hernandez AV, Marti KM, Roman YM. Meta-analysis. CHEST. 2020;158:S97–S102.
10. Systematic review and meta-analysis methodology | Blood | American Society of Hematology. https://ashpublications.org/blood/article/116/17/3140/27947/Systematic-review-and-meta-analysis-methodology
11. Mikolajewicz N, Komarova SV. Meta-analytic methodology for basic research: a practical guide. Front Physiol. 2019;10:203.
12. Andrade C. Mean difference, standardized mean difference (SMD), and their use in meta-analysis: as simple as it gets. J Clin Psychiatry. 2020;81:11349.
13. Chapter 6: Choosing effect measures and computing estimates of effect | Cochrane Training. https://training.cochrane.org/handbook/current/chapter-06#section-6-8.
14. Aguinis H, Dalton DR, Bosco FA, Pierce CA, Dalton CM. Meta-analytic choices and judgment calls: implications for theory building and testing, obtained effect sizes, and scholarly impact. J Manag. 2011;37:5–38.
15. Standardized or simple effect size: What should be reported? – Baguley --2009 – British Journal of Psychology - Wiley Online Library. https://bpspsychub.onlinelibrary.wiley.com/doi/10.1348/000712608X377117.
16. Meta-analysis of data from animal studies: A practical guide - ScienceDirect. https://www.sciencedirect.com/science/article/pii/S016502701300321X?via%3Dihub.
17. Borenstein M. Effect sizes for continuous data. In: *The handbook of research synthesis and meta-analysis*. 2nd ed. New York, NY, US: Russell Sage Foundation; 2009. p. 221–35.
18. Lipsey, M. W. & Wilson, D. B. Practical meta-analysis. ix, 247 (Sage Publications, Inc, Thousand Oaks, CA, US, 2001).
19. Comer, C. Research guides: systematic reviews and meta-analyses: data extraction. https://guides.lib.vt.edu/SRMA/extraction.
20. Harrod, T. Research guides: study design 101: meta-analysis. https://guides.himmelfarb.gwu.edu/studydesign101/metaanalysis.
21. Meta-Analysis Accelerator | Your Meta-Analysis Conversion Tool. https://meta-converter.com/.
22. Abbas A, Hefnawy MT, Negida A. Meta-analysis accelerator: a comprehensive tool for statistical data conversion in systematic reviews with meta-analysis. BMC Med Res Methodol. 2024;24:243.
23. Chapter 10: Analysing data and undertaking meta-analyses. https://training.cochrane.org/handbook/current/chapter-10.
24. Stanimirova I, Walczak B. Classification of data with missing elements and outliers. Talanta. 2008;76:602–9.
25. Elbourne DR, et al. Meta-analyses involving cross-over trials: methodological issues. Int J Epidemiol. 2002;31:140–9.
26. Statistical data preparation: management of missing values and outliers. https://ekja.org/journal/view.php?doi=10.4097/kjae.2017.70.4.407.
27. Analysing data and undertaking meta-analyses - Cochrane Handbook for Systematic Reviews of Interventions – Wiley Online Library. https://onlinelibrary.wiley.com/doi/10.1002/9781119536604.ch10.
28. Dealing With Missing Data – Sainani – 2015 – PM&R – Wiley Online Library. https://onlinelibrary.wiley.com/doi/10.1016/j.pmrj.2015.07.011.
29. Handling trial participants with missing outcome data when conducting a meta-analysis: a systematic survey of proposed approaches | Systematic Reviews | Full Text. https://systematicreviewsjournal.biomedcentral.com/articles/10.1186/s13643-015-0083-6.

30. Higgins JPT, White IR, Wood AM Imputation methods for missing outcome data in meta-analysis of clinical trials. 2008. https://journals.sagepub.com/doi/10.1177/1740774508091600.
31. Ebrahim S, et al. Addressing continuous data for participants excluded from trial analysis: a guide for systematic reviewers. J Clin Epidemiol. 2013;66:1014–1021.e1.
32. Ebrahim S, et al. Addressing continuous data measured with different instruments for participants excluded from trial analysis: a guide for systematic reviewers. J Clin Epidemiol. 2014;67:560–70.
33. Akl EA, et al. Three challenges described for identifying participants with missing data in trials reports, and potential solutions suggested to systematic reviewers. J Clin Epidemiol. 2016;76:147–54.
34. Schwarzer G, Carpenter JR, Rücker G. Fixed effect and random effects meta-analysis. In: Schwarzer G, Carpenter JR, Rücker G, editors. Meta-Analysis with R. Cham: Springer; 2015. p. 21–53. https://doi.org/10.1007/978-3-319-21416-0_2.
35. The Generic Inverse Variance method. *studylib.net* https://studylib.net/doc/5837673/the-generic-inverse-variance-method.
36. Zhai C, Guyatt G. Fixed-effect and random-effects models in meta-analysis. Chin Med J. 2024;137:1–4.
37. Borenstein M, Hedges LV, Higgins JPT, Rothstein HR. A basic introduction to fixed-effect and random-effects models for meta-analysis. Res Synth Methods. 2010;1:97–111.
38. A basic introduction to fixed-effect and random-effects models for meta-analysis – Borenstein – 2010 – Research Synthesis Methods – Wiley Online Library. https://onlinelibrary.wiley.com/doi/10.1002/jrsm.12.
39. Interpretation of random effects meta-analyses | The BMJ. https://www.bmj.com/content/342/bmj.d549.
40. Summing up evidence: one answer is not always enough – The Lancet. https://www.thelancet.com/journals/lancet/article/PIIS0140-6736(97)08468-7/abstract.
41. Introduction to systematic review and meta-analysis. https://ekja.org/journal/view.php?doi=10.4097/kjae.2018.71.2.103.
42. Chapter 23: Including variants on randomized trials | Cochrane Training. https://training.cochrane.org/handbook/current/chapter-23.
43. Mantel N, Haenszel W. Statistical aspects of the analysis of data from retrospective studies of disease. JNCI J Natl Cancer Inst. 1959;22:719–48.
44. Greenland S, Robins JM. Estimation of a common effect parameter from sparse follow-up data. Biometrics. 1985;41:55–68.
45. Beta blockade during and after myocardial infarction: an overview of the randomized trials – PubMed. https://pubmed.ncbi.nlm.nih.gov/2858114/.
46. Much ado about nothing: a comparison of the performance of meta-analytical methods with rare events – PubMed. https://pubmed.ncbi.nlm.nih.gov/16596572/.
47. Issues in the selection of a summary statistic for meta-analysis of clinical trials with binary outcomes – Deeks – 2002 – Statistics in Medicine – Wiley Online Library. https://onlinelibrary.wiley.com/doi/10.1002/sim.1188?msockid=2256bb5171a4615709a1afb570416005.
48. Heterogeneity and statistical significance in meta-analysis: an empirical study of 125 meta-analyses. | Semantic Scholar. https://www.semanticscholar.org/paper/Heterogeneity-and-statistical-significance-in-an-of-Engels-Schmid/42b7610372ec4908f3a146238de70dc3d536b531.
49. Why add anything to nothing? The arcsine difference as a measure of treatment effect in meta-analysis with zero cells – Rücker – 2009 – Statistics in Medicine - Wiley Online Library. https://onlinelibrary.wiley.com/doi/10.1002/sim.3511?msockid=2256bb5171a4615709a1afb570416005.
50. Sinclair JC, Bracken MB. Clinically useful measures of effect in binary analyses of randomized trials. J Clin Epidemiol. 1994;47:881–9.
51. Efthimiou O. Practical guide to the meta-analysis of rare events. Evid Based Ment Health. 2018;21:72–6.

52. What to add to nothing? Use and avoidance of continuity corrections in meta-analysis of sparse data – PubMed. https://pubmed.ncbi.nlm.nih.gov/15116347/.
53. Bradburn MJ, Deeks JJ, Berlin JA, Russell Localio A. Much ado about nothing: a comparison of the performance of meta-analytical methods with rare events. Stat Med. 2007;26:53–77.
54. Meta-analysis of incidence rate data in the presence of zero events | BMC Medical Research Methodology | Full Text. https://bmcmedresmethodol.biomedcentral.com/articles/10.1186/s12874-015-0031-0.
55. Hopkins WG, Rowlands DS. Standardization and other approaches to meta-analyze differences in means. Stat Med. 2024;43:3092–108.
56. 9.4.5.2 Meta-analysis of change scores. https://handbook-5-1.cochrane.org/chapter_9/9_4_5_2_meta_analysis_of_change_scores.htm.
57. Fu R, Holmer HK. Change score or followup score? an empirical evaluation of the impact of choice of mean difference estimates. Rockville (MD): Agency for Healthcare Research and Quality (US); 2015.
58. da Costa BR, et al. Combining follow-up and change data is valid in meta-analyses of continuous outcomes: a meta-epidemiological study. J Clin Epidemiol. 2013;66:847–55.
59. 9.4.6 Combining dichotomous and continuous outcomes. https://handbook-5-1.cochrane.org/chapter_9/9_4_6_combining_dichotomous_and_continuous_outcomes.htm.
60. Expressing findings from meta-analyses of continuous outcomes in terms of risks – PubMed. https://pubmed.ncbi.nlm.nih.gov/21826697/.
61. Chinn S. A simple method for converting an odds ratio to effect size for use in meta-analysis. Stat Med. 2000;19:3127–31.
62. 9.4.7 Meta-analysis of ordinal outcomes and measurement scales. https://handbook-5-1.cochrane.org/chapter_9/9_4_7_meta_analysis_of_ordinal_outcomes_and_measurement_scales.htm.
63. An Introduction to Categorical Data Analysis | Wiley Series in Probability and Statistics. https://onlinelibrary.wiley.com/doi/book/10.1002/0470114754?msockid=2256bb5171a4615709a1afb570416005.
64. Whitehead A, Jones NM. A meta-analysis of clinical trials involving different classifications of response into ordered categories. Stat Med. 1994;13:2503–15.
65. 9.4.8 Meta-analysis of counts and rates. https://handbook-5-1.cochrane.org/chapter_9/9_4_8_meta_analysis_of_counts_and_rates.htm.
66. Hasselblad V, McCrory DC. Meta-analytic tools for medical decision making: a practical guide. Med Decis Mak Int J Soc Med Decis Mak. 1995;15:81–96.
67. Antoniou GA, Antoniou SA, Smith CT. A guide on meta-analysis of time-to-event outcomes using aggregate data in vascular and endovascular surgery. J Vasc Surg. 2020;71:1002–5.
68. Meta-analysis of time-to-event data: a comparison of two-stage methods – Simmonds – 2011 – Research Synthesis Methods – Wiley Online Library. https://onlinelibrary.wiley.com/doi/10.1002/jrsm.44?msockid=2256bb5171a4615709a1afb570416005.
69. Ried K. Interpreting and understanding meta-analysis graphs–a practical guide. Aust Fam Physician. 2006;35:635–8.
70. Elsaeidy AS, et al. Efficacy and safety of extracorporeal membrane oxygenation for cardiogenic shock complicating myocardial infarction: a systematic review and meta-analysis. BMC Cardiovasc Disord. 2024;24:362.
71. Mathew JL. Systematic reviews and meta-analysis: a guide for beginners. Indian Pediatr. 2022;59:320–30.
72. Borenstein M, Hedges LV, Higgins JPT, Rothstein HR. Introduction to meta-analysis. Wiley; 2009. https://doi.org/10.1002/9780470743386.
73. Thiele H, et al. Extracorporeal life support in infarct-related cardiogenic shock. N Engl J Med. 2023;389:1286–97.
74. Lewis S. Forest plots: trying to see the wood and the trees. BMJ. 2001;322:1479–80.
75. Elsaeidy AS, et al. The efficacy and safety of Levosimendan in patients with advanced heart failure: an updated meta-analysis of randomized controlled trials. Am J Cardiovasc Drugs. 2024; https://doi.org/10.1007/s40256-024-00675-z.

76. Melsen WG, Bootsma MCJ, Rovers MM, Bonten MJM. The effects of clinical and statistical heterogeneity on the predictive values of results from meta-analyses. Clin Microbiol Infect. 2014;20:123–9.
77. Higgins JPT, Thompson SG. Quantifying heterogeneity in a meta-analysis. Stat Med. 2002;21:1539–58.
78. Ruppar T. Meta-analysis: how to quantify and explain heterogeneity? Eur J Cardiovasc Nurs. 2020;19:646–52.
79. Shaheen RS, Ismail RA, Salama EY, Korini SM, Elsaeidy AS. Efficacy and safety of 12-hour versus 24-hour magnesium sulfate in management of patients with pre-eclampsia and eclampsia: a systematic review and meta-analysis. BMC Womens Health. 2024;24:421.
80. Page MJ, et al. PRISMA 2020 explanation and elaboration: updated guidance and exemplars for reporting systematic reviews. BMJ. 2021;372:n160.
81. Elsaeidy AS, et al. Efficacy and safety of ketamine-Dexmedetomidine versus ketamine-Propofol combination for Periprocedural sedation: a systematic review and meta-analysis. Curr Pain Headache Rep. 2024;28:211–27.
82. Suvorov AY, et al. Basic aspects of meta-analysis. Part 1. Sechenov Med J. 2023;14:4–14.
83. Glisic M, et al. A 7-step guideline for qualitative synthesis and meta-analysis of observational studies in health sciences. Public Health Rev. 2023;44:1605454.
84. Altenberger J, et al. Efficacy and safety of the pulsed infusions of levosimendan in outpatients with advanced heart failure (LevoRep) study: a multicentre randomized trial. Eur J Heart Fail. 2014;16:898–906.
85. Sharpe D. Of apples and oranges, file drawers and garbage: why validity issues in meta-analysis will not go away. Clin Psychol Rev. 1997;17:881–901.
86. Ioannidis JPA, et al. Comparison of evidence of treatment effects in randomized and nonrandomized studies. JAMA. 2001;286:821–30.
87. The Promise and Problems of Meta-Analysis | New England Journal of Medicine. https://www.nejm.org/doi/10.1056/NEJM199708213370810?url_ver=Z39.88-2003&rfr_id=ori:rid:crossref.org&rfr_dat=cr_pub%20%200pubmed.
88. How should meta-regression analyses be undertaken and interpreted? – Thompson – 2002 – Statistics in Medicine – Wiley Online Library. https://onlinelibrary.wiley.com/doi/10.1002/sim.1187.
89. Goodman S, Dickersin K. Metabias: a challenge for comparative effectiveness research. Ann Intern Med. 2011;155:61–2.
90. Dickersin K. The existence of publication bias and risk factors for its occurrence. JAMA. 1990;263:1385–9.
91. Turner EH, Matthews AM, Linardatos E, Tell RA, Rosenthal R. Selective publication of antidepressant trials and its influence on apparent efficacy. N Engl J Med. 2008;358:252–60.
92. Bias in meta-analysis detected by a simple, graphical test | The BMJ. https://www.bmj.com/content/315/7109/629.
93. Recommendations for examining and interpreting funnel plot asymmetry in meta-analyses of randomised controlled trials | The BMJ. https://www.bmj.com/content/343/bmj.d4002.
94. Bassler D, et al. Stopping randomized trials early for benefit and estimation of treatment effects: systematic review and meta-regression analysis. JAMA. 2010;303:1180–7.
95. Dechartres A, Boutron I, Trinquart L, Charles P, Ravaud P. Single-center trials show larger treatment effects than multicenter trials: evidence from a meta-epidemiologic study. Ann Intern Med. 2011;155:39–51.
96. Afonso J, Ramirez-Campillo R, Clemente FM, Büttner FC, Andrade R. The perils of misinterpreting and misusing "publication bias" in meta-analyses: an education review on funnel plot-based methods. Sports Med. 2024;54:257–69.

How to Recruit a Research Team? 16

Mariam Elgabry, Islam Mohammad Shehata ⓘ, Omar Viswanath ⓘ, and Ivan Urits

> *Alone we can do so little; together we can do so much.*
>
> —Helen Keller

I started my journey of writing in 2019 as a solo writer. I was proud because I had been the one who created the entire article from the idea. However, during my journey, I had noticed that one article could take a couple of weeks just to find a start and much more to finish it. After my eighth paper, I figured out that I must recruit a team, and I was so late. With my team, I have been finding everything much easier. I found innovative ideas, passion, help and support; furthermore, I realized that I could run multiple projects at the same time. On top of that, it has been very inspiring for me to give a hand to someone starting their own journey of writing (Fig. 16.1).

This and more are what made us believe in the TEAM! – Islam Mohammad Shehata

---

M. Elgabry (✉)
Faculty of Medicine, Modern University for Technology and Information, Cairo, Egypt
e-mail: maryam.104330@medicine.mti.edu.eg

I. M. Shehata
Department of Anesthesiology, Faculty of Medicine, Ain Shams University Cairo, Cairo, Egypt
e-mail: Islam.shehata@med.asu.edu.eg

O. Viswanath
Department of Anesthesiology, Creighton University School of Medicine, Phoenix, AZ, USA

Mountain View Headache and Spine Institute, Phoenix, AZ, USA

I. Urits
Southcoast Health Physicians Group, Brain and Spine Center, Wareham, MA, USA

© The Author(s), under exclusive license to Springer Nature Switzerland AG 2025
I. M. Shehata, O. Viswanath (eds.), *How to Successfully Publish a Manuscript*,
https://doi.org/10.1007/978-3-031-92538-2_16

Fig. 16.1 Team benefits

16.1 Team Benefits

16.1.1 Team Recruitment

16.1.1.1 Who?

The leader (typically the first or last author) is the one who recruits the team and influences them to achieve a common goal.

Your work as a leader starts as early as **choosing your coauthors** (Figs. 16.2 and 16.3).

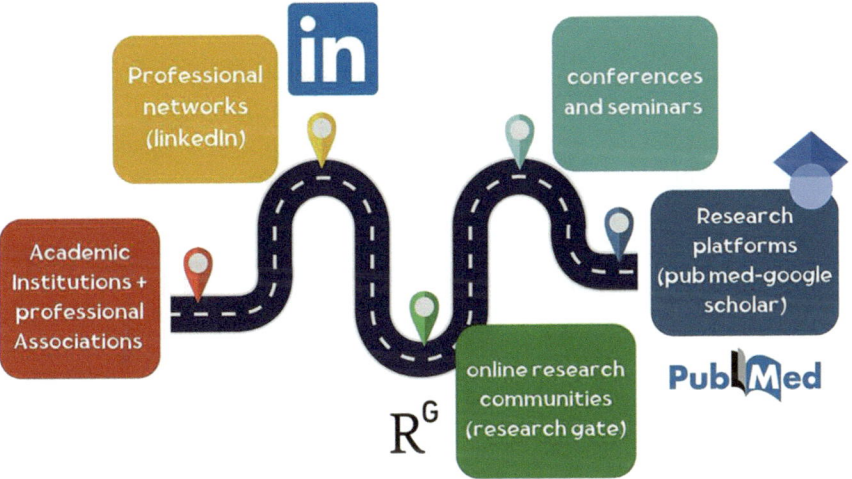

Fig. 16.2 Where to find your coauthors?

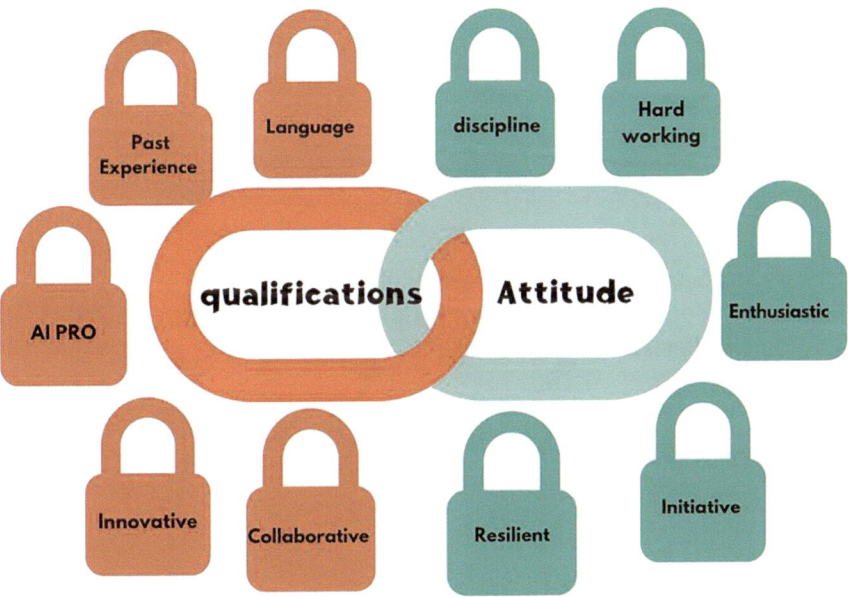

Fig. 16.3 What to look for in your coauthors?

16.2 Where to Find Your Coauthors?

16.3 What to Look for in Your Coauthors?

16.4 Team Roles

Depending on the type of research you are writing, the roles in your team will be decided, as not every research type requires the same specialties (Fig. 16.4).

This article shows an example of the different team roles responsible for the article (Fig. 16.5).

Fig. 16.4 Team roles

Fig. 16.5 Team role's example [1]

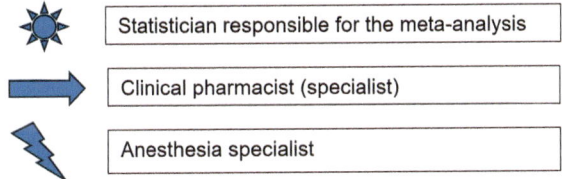

16.5 The Leader's Role [2]

- **"Maintaining the rhythm"** at a timely manner.
- Your responsibilities as a leader:
  - Coordinate project activities.
  - Ensure effective communication.
  - Task assignments.
  - Financial reporting.
  - Setting deadlines.

16.6 Common Problems

Conflict and misunderstanding are normal part of the research process when involving several coauthors, each of whom fulfill specific roles. As a leader, it is your responsibility to resolve any issues that may arise within the research team.

Think about the following scenarios and how you would manage them:

Scenario 1: Conflict between members
1. Set a private meeting with the members involved.
2. Be strict in making the decision.
3. Set guidelines to prevent similar situations.

Scenario 2: Missed deadlines
Prevention
1. Send gentle reminders 1 week/3 days/1 day before the deadline.
2. Set an early deadline.

How to treat
1. Ask about the reason.
2. Offer support and get help from another member.

3. Set new deadlines if possible.
4. If they exceeded the guidelines again, the results would be up to termination.

Scenario 3: Lack of participation in meetings
Prevention
1. Make a safe environment for everyone to express their opinion.
2. Ask about everyone's ideas at the end of every meeting.
3. Direct questions to the quiet member.
4. Assign them to write the draft of the meeting.

Scenario 4: Ethical concerns
Ethical concerns are not negotiable.
– Warn them personally, and if repeated, remove them (because this affects the whole team).

Scenario 5: Low motivation
Prevention
1. Send frequent motivational group messages.
2. Celebrate your team's achievements.
3. Acknowledge your teammates' efforts publicly.
4. For every task, assign a leading member with a supporting member.

How to treat
1. Break down the project into smaller tasks to create a sense of accomplishment.
2. Recruit new members to the team.
3. Discuss the potential benefits of the project, for example, potential publishers.

Now that you have assembled your research team and have strategies in place to lead the project, how do you bring it all together into one cohesive writing group?

16.7 Collaborative Writing Strategies

16.7.1 Definition

An interactive and social process that involves a team focused on a common objective that negotiates, coordinates, and communicates during the creation of a common document [2].

16.7.2 Strategies

Collaborative writing strategies are numerous, but there are four that are the most common [3].

1. **Single-Author Writing (Fig. 16.6)**

single author writing

Fig. 16.6 Single-author writing, collaborative writing

Sequential writing

Fig. 16.7 Sequential writing, collaborative writing

☺ It is appropriate when the writing task is simple.

☺ The writing style would be consistent, so that it may not clearly represent the group's intention.

2. **Sequential Writing (Fig. 16.7)**

   ☺ You will start writing and pass it to the next person to complete their part.

   ☺ Works with poor structure and coordination when it is difficult to meet often.

   ☹ One-person bottleneck (it may invalidate the previous work), "lack of consensus."

3. **Parallel Writing (Simultaneous) Types**
   1. **Horizontal Division (Fig. 16.8)**

      ☺ Effective when a high volume of rapid input is needed.

      ☹ Stylistic inconsistencies because writers may be unaware of each other's work.

2. **Stratified Division (Fig. 16.9)**

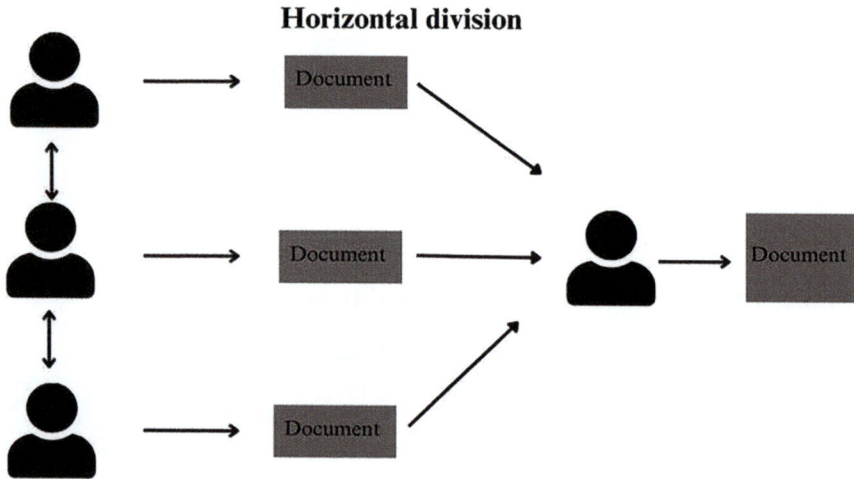

Fig. 16.8 Horizontal division, collaborative writing

Fig. 16.9 Stratified division, collaborative writing

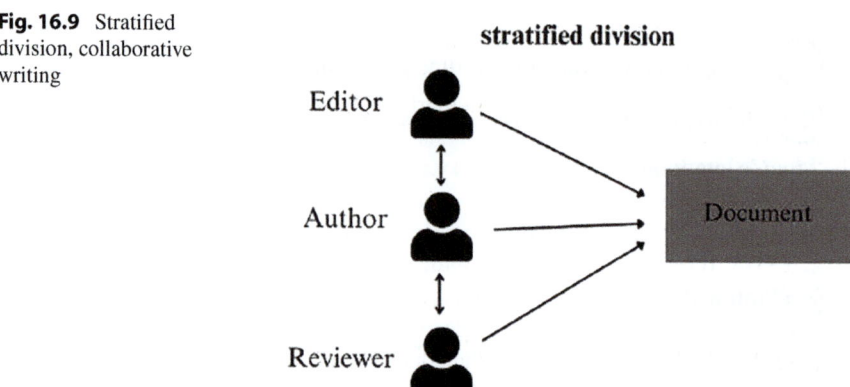

😊 Writing tasks that are difficult to segment and complicated.
😊 This method leverages the varied talents of your group members.
☹️ Takes longer than normal.

3. **Reactive Writing (Fig. 16.10)**

Imagine you write the initial draft and then send it to your coauthors for simultaneous review. They might edit it at the same time, and their changes could align or conflict with yours:

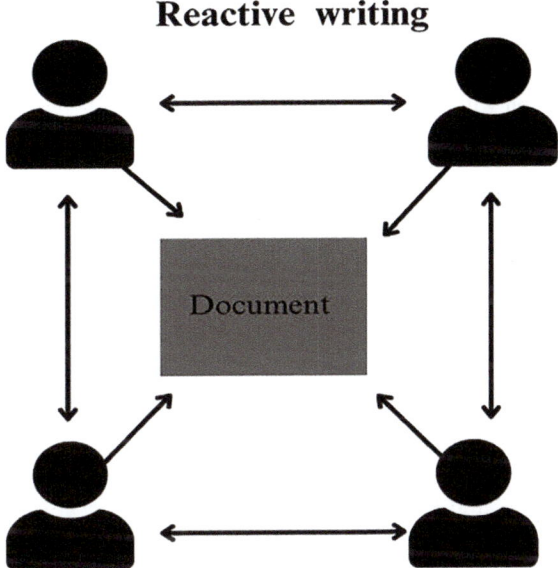

Fig. 16.10 Reactive writing, collaborative writing

☐ All-in reaction, it encourages consensus through dynamic and creative input from all writers.

☐ A powerful tool for interdisciplinary teams to generate innovative ideas that extend beyond traditional disciplinary boundaries.

☹ Potential chaotic development due to impulsive editing.

☹ Seems turbulent for the less experienced teams.

16.8 The Post-article Blues [4]

Now, your first publication has been published! CONGRATULATIONS! However:

You and your team may feel unsure of the next steps since your primary goal has been accomplished—this is called the **post-event blues.**

How to treat:

1. Celebrating success.
2. Accepting downtime and taking some time off for a mental rest.
3. Providing opportunities for skill building and promoting continuous learning.
4. Providing constructive feedback.
5. Setting new goals—use the success from this project to fuel motivation to continue future research endeavors. Success in research and publications can be addicting!

As a part of setting your new goals, you will have to send emails to potential collaborators.

16.9 How to Draft an Email to the Professor

A well-crafted email not only reflects your professionalism but also helps establish a strong connection with potential collaborators (Fig. 16.11).

Here is an easy step-by-step guide:

EMAIL DRAFT

Esteemed professor,

Introduce yourself briefly, mentioning your current position, field of study, , reason for reaching out and your research accounts (ORCID , Google Scholar, PubMed, and Research gate)

Demonstrate **genuine interest in their work** by referencing specific publications or projects and referring to mutual connections between you

Present Your Value to the project and **Request a Next Step** by a clear and polite call-to-action

Thank them for their time and consideration

Add your **name and affiliation**

Fig. 16.11 Email drafting

1. Address the professor formally.
2. Introduce yourself briefly.
3. Demonstrate genuine interest in their work.
4. Explain how your skills or interests align with their work.
5. Conclude with a clear and polite call-to-action.
6. Thank them for their time and consideration.

16.10 Important Points Not to Miss When Drafting Your Email

1. Be concise.
2. Personalize the content.
3. Maintain a professional tone.
4. Proofread thoroughly.
5. Follow up politely.

References

1. Elsaeidy AS, Ahmad HM, Kohaf NA, Aboutaleb A, Kumar D, Elsaeidy KS, Mohamed OS, Kaye AD, Shehata IM. Efficacy and safety of ketamine-dexmedetomidine versus ketamine-propofol combination for periprocedural sedation: a systematic review and meta-analysis. Curr Pain Headache Rep. 2024;28(4):211–27. https://doi.org/10.1007/s11916-023-01208-0.
2. Bolden R. What is leadership? Centre for Leadership Studies, University of Exeter; 2004.
3. Lingard L. Collaborative writing: Strategies and activities for writing productively together. Perspect Med Educ. 2021;10(3):163–6. Available from: https://doi.org/10.1007/s40037-021-00656-4.
4. Davenport Psychology. Post-holiday blues: What are they and what do they mean? [Internet]. 2024 Jan 12 [cited 2024 Nov 18]. Available from: https://davenportpsychology.com/2024/01/12/post-holiday-blues-what-are-they-and-what-do-they-mean/

AI in Research

Hamada Hamdy Elbana, Moataz Maher Emara,
Abdelrahman M. Saad, Islam Mohammad Shehata,
Omar Viswanath, and Mohamed Rehman

17.1 To assist not replaceDefinition

Artificial intelligence (AI) is a set of technologies that enable computers to perform various advanced functions, including the ability to see, understand, and translate spoken and written language, analyze data, make recommendations, and more [1].

H. H. Elbana (✉)
Faculty of Medicine, Modern University for Technology and Information, Cairo, Egypt
e-mail: Hamada.96299@Medicine.mti.edu.eg

M. M. Emara
Anesthesiology and Intensive Care and Pain Medicine, Mansoura University,
Faculty of Medicine, Mansoura, Egypt
e-mail: mm.emara@mans.edu.eg

A. M. Saad
Alexandria University, Faculty of Medicine, Alexandria, Egypt
e-mail: abdelrahman.mohamed2007@alexmed.edu.eg

I. M. Shehata
Department of Anesthesiology, Faculty of Medicine, Ain Shams University Cairo,
Cairo, Egypt
e-mail: islam.shehata@med.asu.edu.eg

O. Viswanath
Department of Anesthesiology, Creighton University School of Medicine, Phoenix, AZ, USA

Mountain View Headache and Spine Institute, Phoenix, AZ, USA

M. Rehman
Eric Kobren Professor, Professor of Anesthesiology, Critical Care, Medicine, Pediatrics,
and Medicine John Hopkins School of Medicine, Baltimore, MD, USA
e-mail: rehman@jhmi.edu

17.2 Introduction

Artificial intelligence (AI) has transitioned from science fiction to the heart of cutting-edge research, offering unparalleled opportunities to enhance the medical research process. As researchers grapple with complex datasets, time-intensive tasks, and the need for precision, AI emerges as a powerful ally. From automating mundane yet essential processes to providing profound insights from data, AI enables researchers to focus on the aspects of science that demand human creativity and critical thinking. With AI as a partner, the potential for discovery expands exponentially [2].

AI is here to stay, not as a replacement for human ingenuity, but as a catalyst for accelerating patient care, research, and discovery. In this chapter, we comprehensively explore how AI can become an ally in your research endeavors. From posing the right research question to refining and presenting your findings, we will show you how to harness AI as your friendly neighborhood assistant, empowering you to make meaningful contributions to medical science.

17.3 History (Fig. 17.1) [3–9]

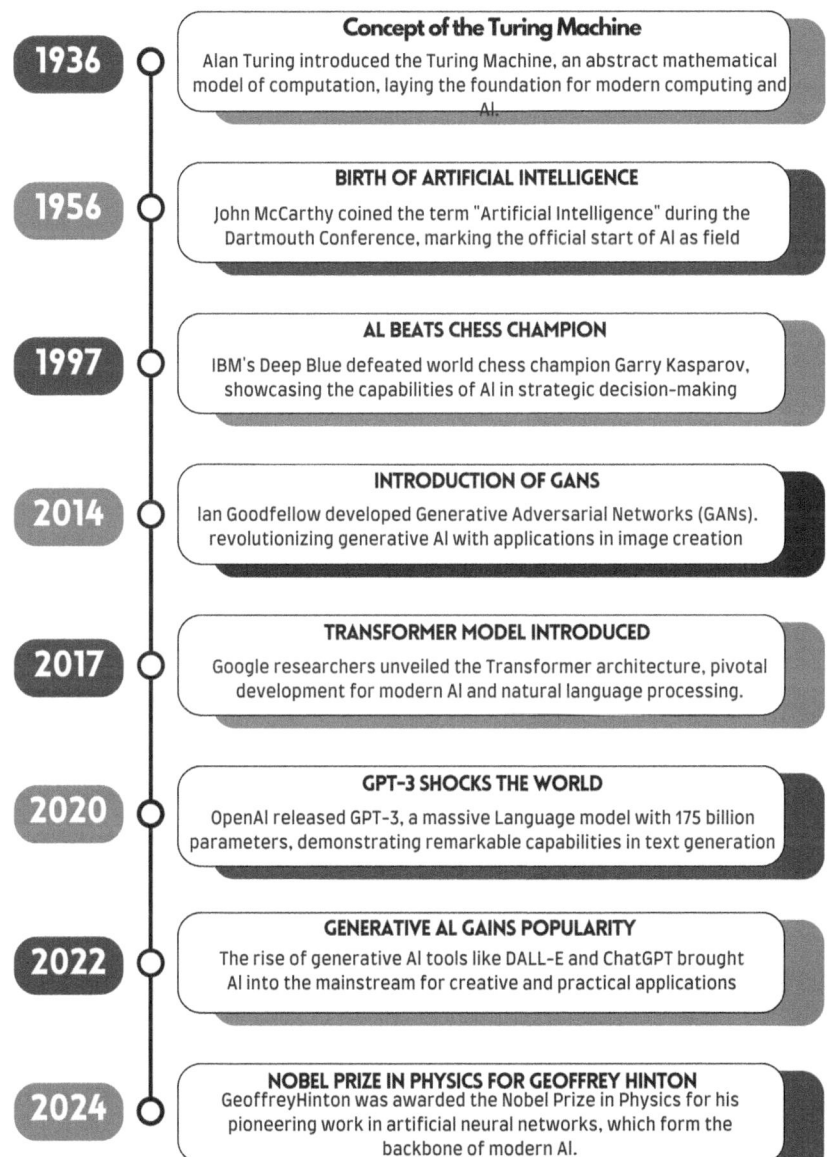

Fig. 17.1 History of AI development

17.4 Importance of AI

Just as AI-powered systems have enhanced clinical settings by assisting clinicians overwhelmed with patient data, they are now poised to be the researchers' indispensable assistant. [10] Imagine a world where AI optimizes literature searches, drafts sections of your manuscript, generates publication-ready figures, and even identifies hidden patterns in your data that could lead to breakthrough findings. This is no longer a distant possibility; it is our reality. [11, 12]

Take, for example, the transformative role of AI in automating repetitive tasks. By saving hours of manual work, AI frees researchers to focus on essential aspects of their projects. [2, 10] Moreover, AI's predictive modeling capabilities are improving clinical trial design and recruitment, enabling faster and more effective studies. [13] These tools have even ventured into uncharted territories, such as discovering new antibiotics or uncovering latent patterns in medical imaging, like predicting sex from retinal fundus photographs—pushing the boundaries of what we thought possible. [14, 15]

17.5 Prompt?

A prompt is the spark given to an AI model to perform a specific action like generating text, photos, or videos.

How prompts (input) are written influences the results (output). Good prompts can result in useful answers, while poor ones lead to bad outputs (Fig. 17.2).

Prompt Engineering
Prompting is the cornerstone used by AI models to generate an output. Natural language processing (NLP) models are all about predicting the next word; so all that the AI model generates depends on your prompt, and this is why there is an interest in prompt engineering. Prompt engineering will change the interaction with AI tools and advance its abilities (Table 17.1).

Fig. 17.2 Prompt importance

Table 17.1 Example: I'm interested in AI application in medicine and research [16]

Prompt	Output
I'm interested in	AI
I'm interested in AI	Application
I'm interested in the AI application	in
I'm interested in AI applications in	medicine
I'm interested in AI applications in medicine	and
I'm interested in AI applications in medicine and	research

17.6 How Do You Generate a Perfect Prompt? [17]

A well-engineered prompt:

- Provides clear instructions.
- Narrows the scope of the response.
- Minimizes ambiguity, leading to more accurate outcomes.

Writing effective prompts for AI tools includes key steps to get the best outputs (Fig. 17.3).

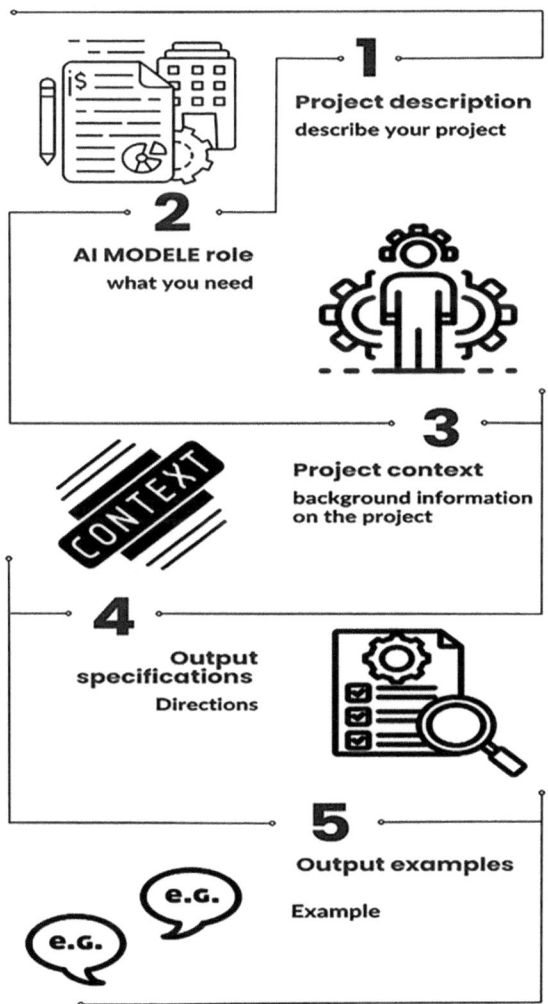

Fig. 17.3 How to write a prompt

1. **Project description**
   Describe your project.
2. **AI model role**
   What do you need from the AI model?
   Example:
   I need you to recommend some titles for this research (mention the type of paper)
3. Project context
   Provide a background on the project or important facts and statistics about it.
4. Output specifications (directions)
   Example "So according to this, I need you to recommend titles for this research (mention the type of your paper) and respect the following criteria (write it guided by the chapter: the title)
5. Output examples
   Provide examples of the output you need. This helps the AI chatbot know better what is in your mind.

17.7 Tips and Tricks for Improving the AI Tool Output (Fig. 17.4)

1. **Use AI tools to generate your prompt as ChatGPT**
   Instead of writing your prompt, you can ask ChatGPT to write it for you.
2. **Use reverse engineering**
   Give the AI tool a sample output and ask it to create a prompt that could lead to that output.
3. **Create your prompt library**
   Every time you use a prompt, save it in your files for later use.
4. **Share prompts with others**
   Tip: Do not overthink the starting prompt. Just initially try and improve it over time.

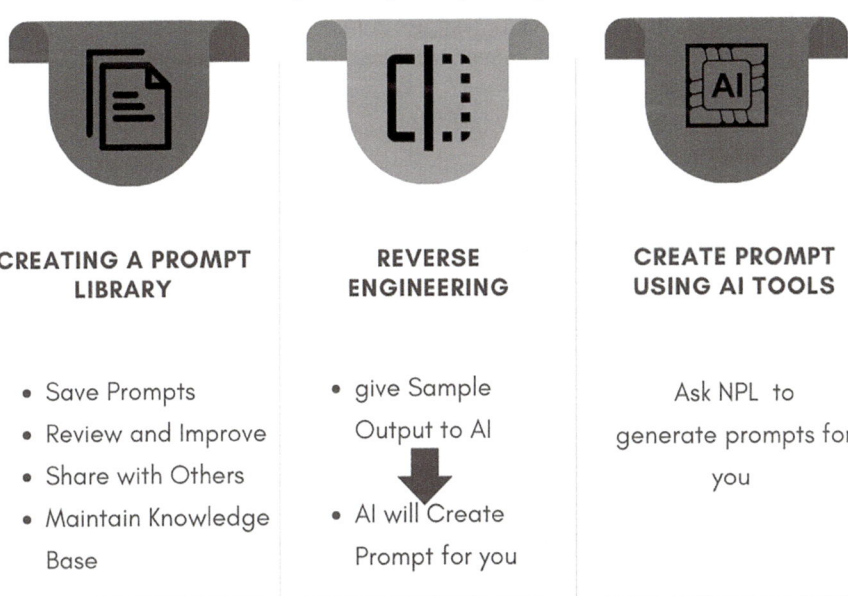

Fig. 17.4 Tips and tricks for improving the AI tool output

17.8 Natural Language Processing (NPL) Tool Supports Scientific Research

There are many NPLs such as **Gemini AI, Microsoft Copilot, Deepseek, and ChatGPT**

The transformative role of AI in automating repetitive tasks: By saving hours of manual work, AI frees researchers to focus on essential aspects of their projects. Moreover, AI's predictive modeling capabilities are improving clinical trial design and recruitment and enabling faster and more effective studies.

What is ChatGPT?

ChatGPT is a natural language processing (NPL) tool that has a great ability to generate human-like responses.

What does GPT stand for?

The GPT in ChatGPT is short for **generative pretrained transformer**. AI training refers to the teaching process of a system to make decisions based on input data.

NPL (ChatGPT) saves time (check our satellites in every chapter)

1. Improving titles and subheadings.
2. Correcting referencing format.

3. Literature review and summarization.
4. Writing and editing: Feed NPL with your introduction or any other part and ask to improve the writing or just give some insights.

   N.B NPL (ChatGPT) can enhance your academic writing; you can use the output as a source of inspiration for writing your paper, but any text generated by AI tools is not allowed to be used (see ethics part below); so use it just for editing or detecting errors.

17.9 AI Hallucinations

Hallucinations are common when using artificial intelligence tools, so you must carefully consider the output results. These are some tips to reduce hallucinations:

1) Avoid unclear or open-ended queries or tasks.
2) AI models hallucinate when tasked with complex tasks.
3) Do not require more than one task at a time.
4) Improve your prompting.
5) Always review the outputs of artificial intelligence tools.

17.10 AI Tools for Every Researcher (Table 17.2)

Table 17.2 AI tools a researcher must have

Tool	Link	Character
R Discovery	AI Academic Research Reading App: AI Literature Search Tool; 250 M+ Research Papers I R Discovery	App to simplify research reading
ExCITATION	ExCITATION — Sort Google Scholar by number of citations, journals' rankings	Find high-quality references in no time
Consensus	Consensus: AI-powered Academic Search Engine	Get insights from the literature
ArXiv Pulse	ArXiv Pulse - Stay updated with the latest research papers	Stay informed with science updates
ResearchFlow	ResearchFlow I AI-Powered Research Engine & Visual Knowledge Mapping	Amplify research with visual mind maps
Resume.io	Job Winning Resume Templates 2024 (Free Download) Resume.io	Create your own resume

17.11 Ethics for Generative AI

Using AI in research is inevitable and valuable, but researchers must follow specific rules and ethical standards to keep the integrity of medical research. Therefore, we will discuss the main ethical principles for using AI-assisted technologies.

1. **Transparency**
   Authors should be transparent about the use of AI in the research process by explicitly disclosing its use in the cover letter, methodology section, and relevant manuscript parts. They should clearly describe how they use AI, the extent of its involvement, and any potential limitations. Thus, it ensures honesty regarding the role of AI in the research and enables reproducibility (Box 17.1).

> **Box 17.1: Guide for Reporting AI Tool Used in Medical Research***
> 1. AI software name.
> 2. Version number.
> 3. Manufacturer.
> 4. Date of use.
> 5. How the AI was used including prompt(s) used and revisions.
>
> This does not apply to basic tools used for checking language and references.
> Authors should confirm their responsibility for the integrity of the AI-generated output.

2. **AI Authorship**
   AI chatbots like ChatGPT **cannot be listed as authors** in medical research because they do not meet the criteria of authorship as per the International Committee of Medical Journal Editors' (ICMJE) guidelines. The four authorship criteria are [18] substantial contributions to the conception or design of the work [19]; drafting or revising the work critically for intellectual content [20]; final approval of the version to be published; and [21] accountability for all aspects of the work, ensuring its accuracy and integrity.
   AI tools cannot perform all these roles, particularly critical thinking, and accountability for the research. As such, researchers should refrain from listing AI as coauthors or citing AI-generated content as standalone contributions, however happened.
3. **Human Accountability**
   Human authors remain fully responsible for revising, editing, and validating the outputs produced by AI technologies, including text, images, citations, etc. AI systems are prone to hallucinations—the generation of inaccurate or fabricated information—making it essential for researchers to critically review and

verify AI-generated content. *Human oversight is not optional; it is a key ethical obligation.*

4. Bias and Fairness

   AI can introduce bias based on their trained data. Researchers must be cautious of potential biases in AI outputs and work actively to mitigate them by analyzing whether the data used to train the AI affects the research's findings. The ethical use of AI requires addressing these biases transparently in your discussion.

5. Data Privacy and Confidentiality

   When using AI, particularly in sensitive fields like medical research, it is crucial to consider the ethical implications of data privacy. Researchers must ensure that AI tools comply with data protection laws like Health Insurance Portability and Accountability Act (HIPAA) and General Data Protection Regulation (GDPR) abbreviation and do not inadvertently compromise confidential or personal data.

6. Intellectual Property and Ownership

   There are evolving questions about the ownership of AI-generated content. Researchers should be aware of the intellectual property implications of using AI tools in research. Who owns the AI-generated content, and how should it be attributed? These questions must be addressed as part of the broader ethical framework for AI use in research.

   Accordingly, authors should revise the copyrights determined by the AI software as a service (SAAS) or owner.

In conclusion, AI tools are aid instruments that assist researchers in conducting and reporting research while also saving time for better, more effective research. When used, it should be declared clearly in the Methods section. Researchers must understand and mitigate their limitations to generate a *trustworthy AI* output. The application of AI in medical research continues to expand; therefore, researchers ought to keep an eye out for new guidelines in this domain.

References

1. What-is-artificial-intelligence. n.d. https://cloud.google.com/learn/what-is-artificial-intelligence
2. Sheikh H, Prins C, Schrijvers E. Artificial intelligence: definition and background. Mission AI: the new system technology. Cham: Springer; 2023. p. 15–41.
3. Muggleton S. Alan Turing and the development of Artificial Intelligence. AIC. 2014;27(1):3–10.
4. Rajaraman V. JohnMcCarthy — Father of artificial intelligence. Reson. 2014;19(3):198–207.
5. Behind Deep Blue: Building the Computer That Defeated the World Chess Champion | Princeton University Press books | IEEE Xplore [Internet]. [cited 2024 Nov 17]. Available from: https://ieeexplore.ieee.org/document/9782447
6. Goodfellow IJ, Pouget-Abadie J, Mirza M, Xu B, Warde-Farley D, Ozair S, et al. Generative Adversarial Networks. arXiv. 2014.
7. Vaswani A, Shazeer N, Parmar N, Uszkoreit J, Jones L, Gomez AN, et al. Attention is all you need. arXiv. 2017.

8. Brown TB, Mann B, Ryder N, Subbiah M, Kaplan J, Dhariwal P, et al. Language models are few-shot learners. arXiv. 2020.
9. Geoffrey Hinton – Facts – 2024 - NobelPrize.org [Internet]. [cited 2024 Nov 17]. Available from: https://www.nobelprize.org/prizes/physics/2024/hinton/facts/
10. Khanna NN, Maindarkar MA, Viswanathan V, Fernandes JFE, Paul S, Bhagawati M, et al. Economics of artificial intelligence in healthcare: diagnosis vs. treatment. Healthcare (Basel). 2022;10(12)
11. Trevisan de Souza VL, Marques BAD, Batagelo HC, Gois JP. A review on Generative Adversarial Networks for image generation. Comput Graph. 2023;114:13–25.
12. Atkinson AG, Lia H, Navarro SM. Advancing scientific writing with artificial intelligence: expanding the research toolkit. Global Surg Educ. 2024;3(1):74.
13. Hutson M. How AI is being used to accelerate clinical trials. Nature. 2024;627(8003):S2–5.
14. Marchant J. Powerful antibiotics discovered using AI. Nature. 2020.
15. Korot E, Pontikos N, Liu X, Wagner SK, Faes L, Huemer J, et al. Predicting sex from retinal fundus photographs using automated deep learning. Sci Rep. 2021;11(1):10286.
16. Generative AI for Everyone. n.d. https://www.coursera.org/learn/generative-ai-for-everyon
17. How To Write ChatGPT Prompts: Your 2024 Guide | Coursera.
18. Flanagin A, Pirracchio R, Khera R, Berkwits M, Hswen Y, Bibbins-Domingo K. Reporting Use of AI in Research and Scholarly Publication-JAMA Network Guidance. JAMA. 2024;331(13):1096–8. https://doi.org/10.1001/jama.2024.3471.
19. https://research-and-innovation.ec.europa.eu/document/2b6cf7e5-36ac-41cb-aab5-0d32050143dc_en. n.d.
20. chrome-extension://efaidnbmnnnibpcajpcglclefindmkaj / https://www.icmje.org/icmje-recommendations.pdf
21. https://research-and-innovation.ec.europa.eu/document/2b6cf7e5-36ac-41cb-aab5-0d32050143dc_en

Index

A
Abstracts
  academic paper, placing, 63
  AI tools, 72
  definition, 62
  Don'ts, 70
  example of, case report, 72
  importance, 62–63
  qualities of, 69–70
  structure of, 64–65
  submission, choosing, 64
  types of, 63–64
  writing, 63, 65
    conclusion, 69
    introduction, 66
    methodology, 66, 67
    results, 67–69
Academic paper, placing, 63
Acceptance rate, 132, 133
Accidental plagiarism, 6
Account creation, 138–140
Acknowledgments, 146
Adapted figures, 78
Al Razi 865 AD, 11
American Psychological Association (APA) style, 118
Annotated bibliography, 112
Argument, construction, 52
Article processing charges (APCs), 154
Artificial intelligence (AI), 266
  authorship, 273
  ChatGPT, 270
  hallucinations, 272
  history, 266–268
  importance, 268
  peer review, 179, 180
  prompt engineering, 268
  research question, 20

Artwork
  definition, 75
  design illustrations, 91, 92
  figures, submitting, 99
  graphs, different types of, 81, 83–87
  importance, 75–77
  photographs, 96, 97
    microscopic images, 97, 98
    radiological images, 97
  preparing figures, considerations in, 78–80
  process diagrams, 88–90
  tables, creating, 80, 81
  types of figures, 77–78
Author waiver, 154
Authorship, 4
  concerns, 4
  definition, 4
  verification, 141

B
Bar graphs, 84–86
Bias, 274
Bibliography, 111–112
Books, 105
Browser connector, 113

C
Canva, 202
Case report
  benefits, 184
  characteristics, 185–186
  consent, 187
  definition, 183
  Don'ts of Writing, 189
  importance, 184
  preparation, 186
  structure of, 187, 188

ChatGPT, 271
ChatPDF, 198
Citations, 109, 111
  definition, 4
  creating, 111
CiteDrive, 111
Clinical heterogeneity, 237
Cochrane library, 202
Cochrane risk of bias tool, 199
Cochran's Q test, 237
Collaborative writing strategies, 258
Color model, 99
Colors, 100
Column graphs, 84, 85
Communication, 257, 258
Comprehensive meta-analysis (CMA), 201
Concluding sentence, 52
Conclusion, 59
  definition, 59
  example for, 61–62
  importance, 59–60
  writing, 60
    Dos and Don'ts, 61
    findings/arguments, summarization, 60
    restating introduction, 60
    significance and implications, 60, 61
Connected Papers, 106
Connected Papers ai, 198
Context, 35
Copyright violation, 8
Counterevidence, 52
Cover letter, 146
  definition, 146
  importance, 147
  structure
    closing remarks, 149
    conflicts of interest, ethical conformity and disclosure, 149
    editor with professionalism, 148
    manuscript details, 148
    novelty and innovation, highlighting, 148
    rationale for, 148
  theme of, 147, 148
Covidence, 196
Crowdfunding, 159

D
Data collection, 196
Data extraction, 193, 197
Data privacy, 274

Databases, searching, 18–20
Descriptive abstract, 63
Digital object identifier (DOI), 108, 151
Direct plagiarism, 6
Discounts, 154
Discussion
  AI tools, 56
  definition, 47
  implications, 56
  limitations, 54
  prepare for, 48
  recommendations, 56
  role of, 47–48
  structure of
    break down your body, 49
    subtitles, examples of, 49, 50
  subtitle, structure of
    argument construction, 52, 53
    citing results, 52–54
    constructing paragraph, 51, 52
    transition words and phrases, organizing discussion using, 54, 55
  writing rules, 48, 49
Duplicate publication, 7

E
EasyBib, 111
Editor, 167
Editorial board, reputation of, 131
Effect size calculation, 200
Elicit, 198
Email draf, 262
EndNote, 113
ENDNOTE 21.4, 115
Epidural anesthesia technique, 94
Established databases, indexing in, 130
Ethics of research, 1
  approval, 9, 10
  authorship, 4
  citations, 4
  considerations, 3
  duplicate publication, 7
  generative AI, 5
  history, 2, 3
  journal submission stage, authors providing at
    acknowledgments, 8
    authors' contribution, 8
    affiliation policy, 9
    data integrity, 8

competing interests/conflict of
interest, 8
funding, 8
pre-submission considerations related
to authorship, 8
plagiarism, 5–7
principles of, 3
role, 1
Exclusion, 196
phase of, 126
avoiding predatory journals, 127
identifying, 128
predatory journals, 126
predatory journals,
characteristics of, 128
target, 127
Experiment, 159

F
Figures
considerations in, 78–80
referencing, 112
resolution, 100
types of, 77–78
Formal email, 262
Funding, 8, 134
open-access models, 154
organization, finding, 156, 157
proposal writing, 157
appendices, 159
background and significance, 159
executive summary scenarios, 158
publication details, 159
title page, 157
writing rules, 157
sources of, 154–156
Funds, 153

G
Galen 130 AD, 10
Generative AI, 5
Generic inverse-variance method
(GIVM), 213
GoFundMe, 159
Google Scholar, 106
Graphs, different types of, 81, 83
bar graphs, 84, 85
column graphs, 84, 85
histograms, 86
line graphs, 83, 84

pie charts, 86
scatter plots, 86, 87
Group labelling, 235

H
Hallucinations, 272
Hand-drawn figures, 100
Harvard referencing, 118
Heterogeneity, 236–238
Heterogeneity assessment, 200
H-index, 122
Hippocrates 375 BCE, 10
Histograms, 86
Human accountability, 273
Hypertrophic chondrocytes, 98
Hypothesis, 14

I
Ibn Sina 980 AD, 11
Images, referencing, 112
Imhotep 2600 BCE, 10
Informative abstract, 63
INPLASY, 194
Institutional support, 155
Intellectual property, 274
Interest statement, conflict of, 144–146
Intermediate sentence, 52
Interpreting meta-analysis, 244
Interquartile range (IQR), 210
In-text citation, 111
Introduction, 31
AI tools, 45–46
example, 34, 43, 44
importance, 32
pitfalls
clear thesis statement, absence
of, 39, 40
existing literature, insufficient review
of, 34, 35
ignoring research gap, 36, 37
inadequate background data, 35, 36
overly technical language, 40–42
poor organization and flow, 42, 43
research significance, failure to
establishment, 38, 39
vague research goals and
purpose, 37, 38
structure, 33
Introductory sentence, 51–52
Inverse-variance meta-analyses, 217

ISBN, 108
ISSN, 108

J
Journal
  choosing, 123, 124
  exclusion, phase, 126
    avoiding predatory journals, 127
    identifying, 128
    predatory journals, characteristics of, 126, 128
    target, 127
  exercise caution upon selecting, 121–123
  potential phase of, 124, 125
    finding, 126, 127
  shortlisting, phase of, 129
    established databases, indexing in, 130, 131
    journal prestige, 129, 130
    journal publication ethics, 132
    publication charges, 133, 134
    publication, timeliness of, 132, 133
    reputation of publisher, 131
  suitable journal, picking, 124, 125
Journal finders, 126
Journal guidelines, 123
Journal impact factor (JIF), 129
Journal publication ethics, 132
Journal submission guidelines, 138
Journal submission stage, authors providing at
  acknowledgments, 8
  affiliation policy, 9
  assessment forms used by, 168–172
  authors' contribution, 8
  competing interests/conflict of interest, 8
  data integrity, 8
  funding, 8
  pre-submission considerations related to authorship, 8

K
Ketamine, 193
Keywords
  choosing, 27
  definition, 26
  demand tool, MESH on, 27
  examples, 27, 28
  ideas, 28
  importance, 27
Kidney, relations of, 95

L
Leave-one-out test, 243
Left ventricular ejection fraction (LVEF), 235
Line graphs, 83–84
Line-numbering, 170
Litmaps, 198

M
Magazines, 105
Managers, referencing, 112, 113
Mantel-Haenszel methods, 218
Manuscript details, 148
Mean difference (MD), 209
Membership fees, 154
Mendeley, 197, 200
Mendeley reference manager, 113
MENDELEY REFERENCE MANAGER V2.125.2, 115
Mendeley web importer, 113
Meta-analyses
  change scores, 227, 228
  continuous outcomes, 225–227
  counts and rates, 230
  data management, 209
  dichotomous outcomes, 218, 220–222, 229
  displaying results, 233, 234
  importance, 206
  limitations, 245
  measurement scales, 229
  missing data, 210, 211
  preparation, 208, 209, 212
  PRISMA, 207
  random effect methods, 217
  rare events, 223
  statistical analysis models, 238
  statistical data conversion, 209, 210
  time-to-event outcomes, 231, 233
  validity, 225
  weighting schemes, 210
Metabias, 246
Meta-regression, 240
Methodological heterogeneity, 237
Microscopic images, 97–99
Minus sign, 107
Mosaic plagiarism, 6

N
Nazi medical war 1939 AD, 11
Newcastle-Ottawa Scale (NOS), 199
Newspapers, 105
Notebook LM, 198
Nuremburg trials 1947 AD, 12

O

Open access (OA) journals, 122
Open-access models, 154
Original figures, 79

P

Parallel writing (simultaneous) types, 259
Pedigree, 91
Peer review, 131, 162, 163
   choosing reviewers, criteria for, 165
   history of, 162, 164
   importance of, 162
   manuscript, reviewers assessment, 164
   reviewers review
      assessment, reviewers' criteria for, 167, 168
      journals, assessment forms used by, 168–172
      manuscript process, 167
      manuscript reading, 166
      outcomes, 172
   reviewers, reply to, 172–174
      AI, 179, 180
      comments and respond, 175–179
      writing, 174, 175
   reviewing, responsible for, 162
   timeliness of, 132
   types of, 165, 166
Peer-reviewed articles, 105
Periodicity, 133
Peto odds ratio method, 218–220
Photographs, 96–97
   microscopic images, 97, 98
   radiological images, 97
Phrases, organizing discussion using, 54, 55
Pie charts, 86–88
Plagiarism, 5, 6
   consequences of, 6
   types of, 6
   websites for detecting percentage of, 7
Post-acceptance phase
   after acceptance, 150, 151
   definition, 150
   digital object identifier, 151
   presentation, 151, 152
   in press, 151
Potential journals, 124, 125
   finding, 126, 127
Predatory journals, 126
   avoiding, 127
   characteristics of, 128
   identifying, 128
Printing charges, 154
Process diagrams, 88–91

Professional societies, 155
Professionalism, editor with, 148
Proposal writing
   funding, 157
      appendices, 159
      background and significance, 159
      executive summary scenarios, 158
      publication details, 159
      title page, 157
      writing rules, 157
PROSPERO, 194
Publication
   charges, 133
      funding, 134
      journal, type of, 133, 134
   fees, 154
   timeliness of, 132
      acceptance rate, 132, 133
      peer-review process, 132
      periodicity, 133
Publisher, reputation of, 131
Publishing agreements, 154
PubMed identifier (PMID), 108

Q

Quotation marks, 106

R

Radiological images, 97, 98
Randomized controlled trials (RCTs), 192
Rayyan, 196
Reactive writing, 260, 261
Reference list, 109
Reference manager application, 113
Referencing, 103
   adding of, 105–112
   citation, 109–112
   format, 112
   history, 103, 104
   importance of, 104–105
   inclusion, 105
   mistakes, 118
   navigating through different sources, 106–108
   organizing references
      additional pointers, 114
      comparison of managers, 114, 116, 117
      reference managers, 112, 113
   refrain from, 105
   sources to, 105
   styles, 109
   using right references, 105
RefWorks, 111

Relative risk, 221, 222, 224, 225
Reprinted figures, 78
Research gap, 36
Research grant, 155
Research question
   artificial intelligence, 20
   characteristics of, 15, 16
   definition, 14
   importance, 14
   searching databases, 18–20
   stepwise approach, 17, 18
   strategies for, 16
Research rabbit, 198
Research registry, 194
Resource locators/ identifiers, 108
RevMan, 201
Risk ratio (RR), 209
RoB 2 Tool, 199
Running title, 26

S
Scatter plots, 86, 87
Scispace, 111, 198
Scite, 198
Secondary referencing, 108
Self-plagiarism, 6, 8
Semantics Scholar, 106
Sensitivity analyses, 207, 212, 246, 248
Sensitivity analysis, 200
Sequential writing, 259
Shortlisting, phase of, 129
   established databases, indexing in, 130, 131
   journal prestige, 129, 130
Single author writing, 258
Standardized mean difference (SMD), 209
Statistical heterogeneity, 237
Stratified division, 260
Structured abstract, 63
Study selection, 202
Sub-group analysis, 238, 239
Submission
   cover letter, 146
      definition, 146
      example, 149–150
      importance, 147
      structure, 148, 149
      theme of, 147, 148
   post-acceptance phase
      after acceptance, 150, 151
      definition, 150
      digital object identifier, 151
      presentation, 151, 152
      in press, 151
   preparing for, 137
      account creation, 138–140
      authorship verification, 141
      checking journal submission guidelines, 138
      uploading files, 140, 141
   title page, preparing
      acknowledgments, 146
      definition, 143
      interest statement, conflict of, 144, 145
      structure, 143, 144
      word count, 146
Submission fees, 154
Subtitle
   examples of, 49, 50
   structure of
      argument construction, 52, 53
      citing results, 52–54
      constructing paragraph, 51, 52
      transition words and phrases, organizing discussion using, 54, 55
Systematic review
   assessing study quality, 198, 199
   data extraction, 197
   data synthesis, 200–202
   definition, 192
   importance, 192
   steps, 193–195

T
Tables, creating, 80, 81
Team benefits, 254
Team recruitment, 254
Team roles, 256
Text citation, 109
Thematic analysis, 200
Title, 23, 80
   AI tools, 26
   to avoid, 25
   examples, 27, 28
   formatting of, 24
   process, 25
   running title, 26
   types of, 24
Title page preparation
   acknowledgments, 146
   definition, 143
   interest statement, conflict of, 144, 145
   structure, 143, 144

word count, 146
Transition words, organizing discussion using, 54, 55
Transparency, 273
Typing, 101

U
Uniform resource locator (URL), 108
Unstructured abstract, 63
Uploading files, 140, 141

V
Validation, 14
Vancouver style, 118

W
Word count, 146
Word processors, 114
Writing
    abstracts, 63, 65
        conclusion, 69
        introduction, 66
        methodology, 66, 67
        results, 67–69
    Do's and Don'ts, 61
    significance and implications, 60

Z
Zotero, 113, 196
ZOTERO 7.0.9, 115

GPSR Compliance

The European Union's (EU) General Product Safety Regulation (GPSR) is a set of rules that requires consumer products to be safe and our obligations to ensure this.

If you have any concerns about our products, you can contact us on ProductSafety@springernature.com

In case Publisher is established outside the EU, the EU authorized representative is:

Springer Nature Customer Service Center GmbH
Europaplatz 3
69115 Heidelberg, Germany

Batch number: 09397563

Printed by Printforce, the Netherlands